Building the Hobby Stock/Street Stock Car

A Beginners Guide to Hobby Class Racing

By Bob Emmons

ISBN No. 0-936834-26-9

Editor Steve Smith
Associate Editor Georgiann Smith
Photos by: Bob Emmons, Ron Buck, Steve Smith, Ace Lane, Jr., Ray Masser, Mike DeSantis, Kris Hauge

Printed and manufactured in United States of America

Copyright © 1979 by Steve Smith Autosports. No part of this publication may be reproduced in any form or by any means without the express written permission of Steve Smith Autosports.

Published by

Steve Smith Autosports

P.O. Box 11631, Santa Ana, CA 92711

STEVE SMITH AUTOSPORTS PUBLICATIONS

Racer's Complete Reference Guide
The "Yellow Pages" of the high performance world. Where to get: Hardware, Chassis/engine parts, running gear, etc. Where to find fabricators, car builders, engine builders, all parts sources, etc. A unique and valuable reference source. #S108..$9.95

The Complete Karting Guide
The complete guide to kart racing for every class from beginner to enduro. Contains: Buying a kart/equipment. Setting up a kart – tires, weight distrib., tire/plug readings, exhaust tuning, gearing, aerodynamics, engine care, competition tips. #S140..$10.95

Racing The Yamaha KT 100-S Engine
Hi-perf. building/blueprinting one of the most popular 2 stroke engines. Includes: 2 cycle engine basics, Blueprinting and tuning, Tips for ultimate power, Building for reliability, Complete engine machining, Ignition system, Carburetor tips, Using alcohol, Expansion chamber and exhaust tuning, Maintenance. #S151..$9.95

Building A Street Stock Step-By-Step
Step-by-step info, from buying the car and roll cage kit to mounting the cage, stripping the car, prep tips, front/rear suspension modifications, adjusting the handling within street stock rules, at-the-track setup, and mounting seats, gauges, linkages, etc. A complete guide for entry level street stock class. #S144..$9.95

The Stock Car Racing Chassis
The basic stock car handling "get acquainted" book. Covers: Under and oversteer, The ideal chassis, Spring rates and principles, Wedge, Camber/caster/toe, Sorting a chassis with tire temperatures, Guide to bldg. a Chevelle/Monte Carlo chassis. #S101..$6.75

So You Want To Go Racing?
A book written for the enthusiast considering his first start in racing. Gives a complete understanding of: Setting up a shop, Buying the right tools/equipment, Organizing/gathering a crew, Getting the most from a limited budget, Being resourceful, Guide to classes of racing, And more. This book is like being an apprentice to a 20-year racing veteran. #S142..$9.95

Trailers - How To Buy & Evaluate
Complete guide to: Buying a used or new trailer, Designing and engineering a trailer, Trailer components, tongues, couplers, the right hitch/tires, Creating towability, Eliminating sway, Trailer brakes, suspension, stability. #S150..$12.95

Race Car Graphics
How to design a race car color schemes, plus numbering, painting and striping, with paint or vinyl. Make your car look professional for a quarter of the cost! Includes 8 pgs of full color. #S118..$8.95

Stalking The Motorsports Sponsor
The methods in this clear, precise book will help you acquire a sponsor and set up a sponsorship program whether you race on a local, regional or national scale. #S119..$8.95

Building A Mini Stock Step-By-Step
From fabricating the tubular frame to fitting the fiberglass body, this book covers it all. Plus a complete hi performance build up of the Ford 2000/2300 cc 4-cyl engine. Suspension, running gear and at-the-track adjustments. #S152..$10.95

Dirt Track Chassis Blueprint
Offset design, left side bias, rear bias, includes full roll cage details. A rigid, lightweight design. Includes details for a 5th coil-over or leaf spring rear suspension. #B500..$17.95

Chevelle Chassis Blueprint .... #B301...$11.50
Square Tube Chassis Blueprint .... #B350...$11.50

Buick Free Spirit Power Manual
Complete info on the Buick V6 performance engine. Blocks, cranks, rods, valve gear, intake and exhaust systems. All new revised edition. Complete list of performance hardware. Over 200 photos and drawings. #S123..$9.95

Racing Engine Preparation
By Waddell Wilson, one of the top NASCAR engine bldrs. He takes you thru all the steps of engine building as carefully as he builds one. Explains: Cam timing/selection, Head porting, Carb mods, Prepping rods/pistons/blocks, HP ignition and oiling systems. "The most complete engine book I've ever seen," says A.J. Foyt. Uses small block Chevy & Ford. #S106..$10.95

Racing The Small Block Chevy
A totally up-to-date hi perf. guide to the small block Chevy. Covers: Hi performance at a reasonable cost, Component blueprinting, Revealing cam, carb, head porting and ignition tips from many of racing's best known engine bldrs. #S112..$10.95

Guide To Optimizing Your Ignition
Most complete book on hi performance automotive ignitions ever written. Conventional and electronic. Includes: Hi performance from factory electronic ignitions, Improving engine/ignition performance, Creating custom tailored systems, Complete troubleshooting guide. Tech tips for ultimate ignitions. #J153..$12.95

Sprint Car Technology
Complete book on buying/building/racing a sprint car, midget or super modified. Covers: Torsion bars, Chassis structure, Straight front axle suspension, Live rear axles, Birdcages, Lateral locating linkages, Steering, Bump steer, Shocks, Dirt and asphalt setups, Tire selection and more. #S125..$11.95

Building The Hobby Stock Car
For the low-buck racer with limited facilities. How to build a car with a minimum of "store bought" parts or expensive machine work. Covers: Car choice, cage construction, Unitized vs. full frame chassis, Engine cooling/electrics, Trans/rear ends, Suspension (parts choice, bldg and sorting), Driving, Preparation. #S126..$9.95

Computerized Chassis Set-Up
This WILL set-up your race car! A computerized analysis of everything in your suspension. Analyze weight and spring rate changes and much more. Plus rating any type of spring, selecting optimum spring rates, cross weight and weight dist., computing CGH, front roll centers, min. rear stagger, ideal ballast location, final gearing and more. For any IBM compatible computer (5.25" or 3.5" disk) or Commodore. Specify your computer type. #C155..$99.95

How To Run A Successful Racing Business
How to operate a racing-oriented business successfully. How to start a business, Put more profit in a business, Cost controls, Proper advertising, Dealing with employees, taxes and accountants, and much more. The secrets of business fundamentals and how to make them work for you. #S143..$9.95

Racer's Guide To Fabricating Shop Eqpmnt
Would you believe an engine stand, hydraulic press, engine hoist, sheet metal brake and motorized flame cutter all for under $400? It's true. Step-by-step instructions. Concise photos and drawings show you how to easily build them yourself. #S145..$10.95

The Racer's Tax Guide
How to save BIG money on your racing activities...even under the newest tax law. Tells how to LEGALLY subtract your racing costs from your income tax by running your operation like a business and following 3 simple steps. An alternative form of funding your racing. Includes a concise up-to-the-minute report on the new tax laws. Read this book now and start saving! #S116..$10.95

ORDER HOTLINE (714) 639-7681
FAX LINE (714) 639-9741

*Add $1.50 per book for 4th Class up to $4.50
● For 1st Class add $3 per book up to $9 ● Allow 3 weeks for 4th class delivery
● In Canada add $3 to all orders ● Checks over 30 must clear first
● Canadian personal checks not accepted ● Calif. residents add 6% tax

Advanced Race Car Suspension
The latest tech info about race car chassis design, set-up and development. Weight transfer, suspension and steering geometry, calculating spring/shock rates, vehicle balance, chassis rigidity, banked track corrections, skid pad testing, and much more. #S105..$9.95 WorkBook for above book: #WB5..$5.95

Super Tuning Holley Carburetors
Tips and details that make carburetors win races for you! Carb basics, high air flow for maximum power, throttle linkage tips, troubleshooting, prepping and modifying metering bodies, throttle bodies, etc. A real power secrets book. #S156..$11.95

Dirt Track Chassis Technology
Covers EVERY facet of dirt track racing and stock car chassis set up. Includes: Left side bias, aerodynamics, leaf vs. coil suspensions, 5th c/o and torque arm suspensions, chassis construction, front and rear suspensions, and more. #S133..$12.95

Race Car Fabrication & Preparation
Includes thorough discussions of Chassis/suspension fabrication, Preparing a trans, Setting up a Ford 9" rear, Electrical system, Bldg the roll cage, Cutting costs and beating economy stock rules, Welding, Clutches, Wrecking yard parts, Front end design, Plumbing, Driveline, Wheels and tires, and more. #S114..$10.95

Stock Car Chassis Technology
All the newest ideas in: Stock car chassis set up principles, Procedures for set up at the shop and track, Actual case studies, Front suspension set up and adjustment, Rear suspension - torque arms/stagger/locating devices, Repairing a crashed car. New aerodynamics for short tracks, and much more. For paved and dirt tracks. #S139..$12.95

Bldg The Pro Stock/Late Model Sportsman
A complete build-up of a low cost chassis for the pro stock/sportsman class. Uses 1970-81 Camaro front stub with fabricated perimeter frame. How to design ideal front and rear roll centers, ideal camber change curves, Flat track vs. Banked track set ups, Effects of spindle drop, Effects of mass placement, rear suspensions for dirt and asphalt, step-by-step chassis set-up, and much more. Includes blueprint for bldg a chassis and cage. For dirt and paved tracks. #S157..$11.95

Bldg The Stock Stub Race Car VIDEO
This new video is a MUST to accompany the above book. Complete step-by-step chassis fabrication. #V164...$39.95 Special Package Price for #S157 & #V164 is only $45.95!

Other VIDEOS
Asphalt Chassis Set-Up by Ray Dillon....#V159...$39.95

Dirt Chassis Set-Up by Charlie Swartz....#V160...$39.95

Engine Maintenance & Power Tuning by Ray Baker.....#V161...$39.95

Short Track Driving Techniques
By Butch Miller, ASA champ. Includes Basic competition driving, Tips for driving traffic, Developing a smooth & consistent style, Defensive driving tactics, Offensive driving techniques, and more. For new and experienced drivers alike. #S165...$9.95

STEVE SMITH AUTOSPORTS PUBLICATIONS

P.O. Box 11631 Santa Ana, CA 92711

Table of Contents

Chapter		Page
Chapter 1:	Introduction	1
	Your Car	2
	Your Engine	3
	Tools	3
	Place To Work	3
	Cost	4
Chapter 2:	Roll Cage Construction	6
	Pipe	6
	Tubing	9
	Cage Kits	9
	Building A Cage	11
	Mounting The Cage	11
	Cage In A Unitized Body	12
	Summary	13
Chapter 3:	Supports In Addition To Main Cage	15
	Rear Suppports	15
	Front Supports	16
	Bumpers	18
	Nerf And Rub Rails	19
	Summary	19
Chapter 4:	Frame Chassis Vs. Unitized Body	22
	Advantage Of A Frame Chassis	22
	Chassis Under Unitized	22
	Interior Sheetmetal	24
	Spindles And Ball Joints	25
	Reworking A-Arms	25
	Shock Absorbers	26
	Leaf Spring Wrap-Up Control	27
	Seats And Belts	27
Chapter 5:	Engine, Cooling, Electrics	31
	Engines	31
	Pans	32
	Radiators	34
	Wiring	36
	Fuel Lines	36
	Fuel Tanks and Cells	37
Chapter 6:	Transmissions and Rears	39
	Ratios	39
	Locking The Rear	40
	Gearing The Car	41
	Coolers	41
	Types Of Rears	41
	Floaters	42
Chapter 7:	Tires And Wheels	43
	Wheels	43
	Using Street Tires For Racing	44
	Cheater Tires	45
	Stagger	46
Chapter 8:	Suspension	47
	Coil Springs And Torsion Bars	49
	Leaf Springs	49
	Wedge Bolts	51
	Anti-Roll Bars	52
	Suspension Linkages	54
Chapter 9:	Driving	55
	Observation	56
	Dealing With Your Fans And Non Fans	57
	Summary	58
Chapter 10:	Preparation And Spare Parts	59
	Spare Parts	62
Chapter 11:	Racing Is Fun, But!!!	64
	Crew	64
	Hangers On	65
	Sponsors	65
	Officials	65
	Dealing With The Public	66
Appendix A:	Common Stock Car Terms	67
Appendix B:	Cage Kit Makers	68
Appendix C:	Other Helpful Sources	68

THANKS

It is impossible to thank all the people who have helped me in racing over the last twenty years for fear of leaving someone out, but a special thanks to Harry T. Treacy who put up with me for a partner in more race cars than we would either care to count.

Robert Emmons

Chapter One

Introduction

The term "Hobby Car" covers a variety of rule classifications depending upon the track you plan to run. These vary from the strictly stock rules that allow almost no modifications—other than for safety—to those which allow extensive and expensive speed shop parts. We will attempt with this book to help those people who are just starting out in a hobby class at a track which limits their class to something less than a "store-bought" late model sportsman car.

Before You Start

Your first move is to get a copy of the rule book for the track or tracks you wish to run, then study it. Also, talk to car owners in your class. Most people flock around the drivers after the races, but you want to talk to the person who built the car. Most people who build cars are glad to talk to you about their creation when approached at the right time and in the right way. Don't try to strike up a conversation when the crew is trying to put the car back together for the consolation race or when two wreckers have just deposited on their trailer the remains of an encounter with the wall.

The more you learn, the more intelligent the questions you can ask. And with more intelligent questions, the more willing the person will be to talk. Listen to everything that is said and then sort out the good ideas from the poor. A side benefit is that when you finally get

Get the rule books you need for the tracks you are going to run. They and some good racing tech books will be your guide.

Go in the pits. You can't learn anything about building a car in the grandstand.

your own car to the track, you'll have someone that might lend you a hand because they know you. Most racing people are more than willing to help beginners as long as they appear serious about racing.

A good investment is a track license and pit pass even before you have a car of your own. The view of the race track may not be as good in the pits as in the grandstand, but you can learn a lot more by looking at the cars up close in the daylight than you can in the dark for the half hour or so you have after the races. When you see an idea you like, write it down or sketch it in a notebook. It is too easy to forget later on when you need it. It is only common courtesy to ask permission first before you go sticking your head in someone else's car. Many people are very possessive of their equipment.

The best way to gain experience for no monetary investment is to hook on with the crew of an existing car even if it is a car that does not run in your class. In this way you can find out much about the operation of keeping a car running without spending your own money. If it is a good crew, you learn how to do things right. If it's a poor operation, you can learn how not to do things. Both will be valuable to you. Try to work with the most competitive car you can find—preferably in the top class at your track. Quite obviously, the people operating this car will be the most experienced, and, they will positively influence you.

Your Car

Since you have already gone to the expense of buying this book, you are fired up to get to work on your car. You can judge by the front runners which cars seem to be the hot setup. Be careful though—especially at tracks with liberal rules. As you may find that sheet metal is only skin deep. The chassis is what makes the car work on a short track, not the body. For a first class car, we would suggest something with a full frame rather than a unitized body. This will be discussed more fully in a following chapter.

All hobby classes have a few high bucks operations, but they aren't necessarily the fastest cars.

If you don't work on a race car, it won't run.

Fords are fun too. But if you are a Mustang or Cougar lover, be sure you can run a full frame under it rather than run it as a unitized car.

Your Engine

Don't get carried away with your motor unless you can afford to keep putting it back together all the time. Many people that come into oval tracking from drag racing are "motor mad". They place too much importance on a powerful engine. Try to avoid this. You will be much better off with a dependable engine even if it is a little down on power. You can't gain experience unless your car is racing. We have seen many cars that were short on power drive around their high buck counterparts in the corners because they had their handling act together. Watch at the next race and you'll see it also.

Tools

Take time to take stock of your supply of tools. You should have the basic wrenches, sockets, screwdrivers and pliers that most people have. As far as tools go, here are three lists to draw from:

Should have (or have access to)

Set of cutting and welding torches
Arc welder (200 amp, should be adequate for race car work)
Tie rod fork
Several hammers including 4-pound short handled model
¼-inch drill and bits
4 jack stands
Set of Allen keys or Allen sockets
Cold chisels and center punch
Variety of files including round and 3-cornered
12 foot tape measure
Torque wrench
Thin 8-foot tape measure (for measuring tires)
Tin snips
Pop rivet gun
Protractor (preferably a magnetic one)

Nice to have

Right-hand and left-hand compound snips
½-inch drill and bits
6-inch and 12-inch adjustable wrenches
Floor jack
Set of line wrenches
5-foot pry bar (or piece of heavy wall pipe)
2 more jack stands
6-pound long handled sledge hammer
Assorted C-clamps
¼-inch drive set
Bench grinder
Body grinder
Cable hoist (come-a-long)

Great to have, but expensive

Drill press
Power hack saw
Metal lathe
Air compressor
Spray gun
Tap and die set

Try to gather tools of your own. It saves time and a lot of hard feelings if you don't have to borrow. Try to keep your tools organized so that you can find them in a hurry. It is not necessary to have a big fancy tool chest to be organized. You can use several of the small tool boxes and accomplish the same thing cheaper. Another advantage of the small boxes is in loading and unloading your tow vehicle. Those big chests are a real hernia maker when you have to handle them alone.

Place To Work

Race cars have been built in dirt driveways so it is possible to work under almost any conditions. If you don't have a garage where you live, there are several things you can do. Perhaps one of your helpers will have some place where you can work or maybe there is a corner where you work. The only problem with this setup is that you can wear out your welcome pretty quickly.

You may rent a place to work, but this can become expensive. A solution that racers in many areas use is to get four or five people together and rent a building. In a situation like this, though, make sure everything is spelled out

This car was built in a driveway because they didn't have a garage. It was also the first car they had ever built from scratch.

Racers are very good at improvising.

Cost

No one can tell you what your car will cost. For a hobby car, this depends upon how well you can scrounge supplies. Another way to keep costs down is to trade. You don't necessarily have to trade parts. No matter what your occupation, there is probably someone out there who can use your services in exchange for work or parts for your race car. Another place to frequent is racing flea markets which have become popular in some parts of the country. The theory behind these is: "What is one person's junk is someone else's gold."

as far as rights and responsibilities beforehand. Otherwise, your friends could become your former friends. Try to get on your own if you can. It saves a lot of problems.

You should have a 220-volt electric service so you can use a welder. Since most car building is done in the Winter months if you live in the northern areas, you're going to need some kind of heat. The small kerosene heaters known as torpedo heaters do a good job for the price, but even a potbelly stove will keep the tools warm enough to handle. While we're talking about heat, *never* store gasoline or any flamables in the garage or use gasoline to wash parts. This is lethal. There are solvents such as Varsol or Parsolv on the market that do a better job and are not nearly as dangerous.

If you tow with a car, you'll need a box like this to carry everything. A trailer will be a very important part of your racing operation.

A single axle trailer is adequate if the hubs and tires are strong enough and the tow vehicle will take the hitch weight. They can get exciting to drive sometimes, because they do not have the stability of a double axle trailer.

A floor jack is almost a necessity, even a heavy old mechanical like this one.

This trailer has much storage space, but would be expensive to build.

If you have a spare vice around, mount it on the trailer. The tire rack on this trailer is removeable so the trailer can be used to haul other things around. This is a good way to make some bucks in your spare time.

This box was built to fit between the rails of the trailer. It has to hold a lot as the owner tows his race car with a motor home.

Everybody reads the local racing papers.

Chapter Two
Roll Cage Construction

There are three basic ways to build a cage. You may fabricate your cage with black iron pipe; you may fabricate it using .095 inch wall tubing, or you may purchase a cage kit.

Pipe

The cheapest method is to use pipe, as many times you can find a supply at reasonable price. (Know someone who works in the plumbing trade?) Make sure that the rules at your track do not specify 1¾-inch outside diameter for the cage. If they do, you will have to use 1½ inch I.D. pipe which is quite heavy (2.731 pounds per foot.) Most people who use pipe use 1¼-inch I.D., which has an O.D. of 1.66 inch (about 1 5/8.) This weighs 2.281 pounds per foot. Make sure you use decent pipe and not some rusty old junk that has been buried in the ground for ten years. Rust causes the walls to become thinner and rust makes a terrible weld. It also weakens the structure of the metal and builds in failure points. Also, never use the threaded part of a piece of pipe for anything. It is too weak. Don't use galvanized pipe. It is a royal pain to weld.

To build your cage from pipe, you will need a pipe bender and some way to notch the pipe so two pieces will mate without a large gap. There are a variety of pipe benders—most of them compression benders—and the mechanics of using the bender will depend on what kind you can borrow. These can be rented from a tool rental outfit, but if you do rent one, do it on the weekend so you can get the fullest use for the price. Chances are it will not go as fast as you plan if you are building a cage for the first time. If you can't borrow or rent a bender, you will have to farm out the work. Ask around the pits or at your local speed shop for someone who might do the work for you. If you do the bending yourself, you will have to do a little experimenting to find out how much set back to use in your bends. By this, it means you must start your bend about seven or eight inches before the final bend. (See drawing.) Most bends are made with about an eight-inch radius. Unless it is a one-shot bender, you will have to take several bites for each bend. You will get small dimples on the inside of the bends. This is normal, but if a wall starts to collapse, cut off and throw away that bend. Pipe has been bent around trees or chained to trailers while being bent with heat from a torch, but usually the results are as crude as the methods.

Once you get the hang of it, bending is not bad at all. Notching is something else again. It is a pain in the neck unless you have the right equipment and chances are you won't. A tubing notcher that is specifically for racers is available at a reasonable price compared to most com-

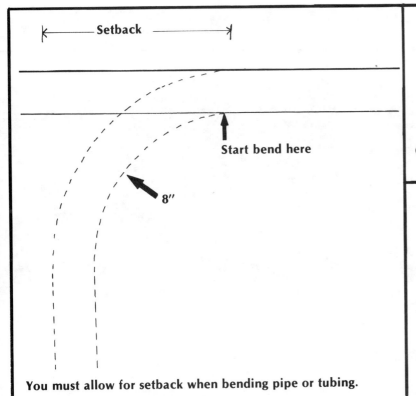

You must allow for setback when bending pipe or tubing.

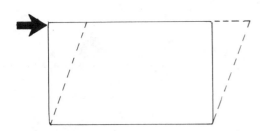

Rectangle (solid lines) is not structurally sound. Force at arrow will cause rectangle to parallel (shown with dotted lines) because verticals are in shear.

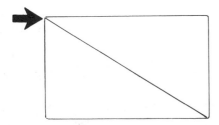

With diagonal added to rectangle you now have two triangles. This is very strong with little weight added as it places all the lines in compression or tension rather than shear.

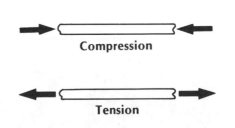

Cage in compression and tension has strength

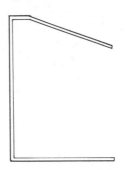

Typical channel type side rail

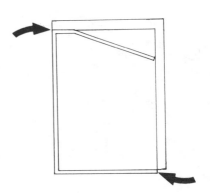

Boxing in channel with angle iron
Weld at arrows

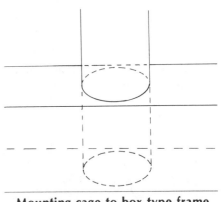

Mounting cage to box type frame

If you tack this cage together in the car and then take it out to weld it, the welding is a lot easier.

Tubing notched at 90 degrees and 45 degrees.

Note how all the bars come together at one point. This is the mark of a well designed cage and support bar structure.

This narrow homemade chassis required outriggers to be welded to the frame rails so the cage had something on which to sit. It would have been easier in the long run to build a perimeter chassis.

mercial notchers, but the price is still prohibitive for building a single car. It does a great job and, if you stay in racing for any length of time, it is a worthwhile investment. Once again, check with your trusty tool rental store. They might have one.

If you have a lathe or milling machine available, you can purchase a cutter of the right diameter and fabricate a holder so that the tubing can be fed into the cutter. If you do this, make sure that you can feed at an angle other than a right angle. Some of your joints will be less than a ninety degree angle. Some people have used hole saws in drill presses to notch. This is slow and the saws tend to dull or break quickly.

If you have a power hack saw or cutoff wheel available, you can do with little effort the same thing that you can do with your hand hack saw. To butt two pipes at right angles to each other, you must cut two 3/8 inch deep, 45-degree cuts on the end of the pipe. (See diagram.) A touch with a grinder and there should be very little gap to fill with weld. If you have too big a gap, keep grinding until it fits. I've seen cages with bolts and nuts used to fill gaps. This not only looks terrible, but is not good welding practice. See photo for notching at other than 90 degrees.

A last method is to notch with a cutting torch and finish with a grinder.

I have used most of the above methods and come up with good looking, safe cages. No matter how little you have to work with, if you want to race, you'll get the job done if you have to buy hack saw blades by the dozen to do it. A cage takes time, effort and patience. If you don't

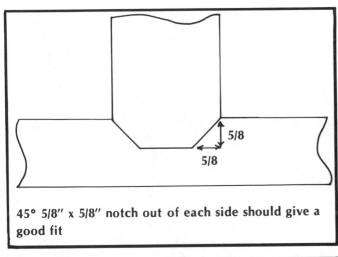

45° 5/8" x 5/8" notch out of each side should give a good fit

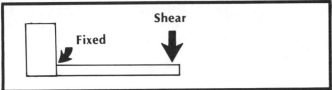

These side bars were made without a pipe bender. It isn't a bad idea, but without uprights and gussetts, these are unsafe.

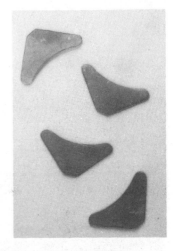

In the long run these store bought gussetts are cheaper than making your own.

have all of them, don't attempt it.

Tubing

Building a cage from steel tubing is similar to working with pipe except you must have access to a tubing bender. These are different from pipe benders in that they do not distort the thin wall tubing and it can retain the structural integrity of the tube. Never use a pipe bender on tubing. It will collapse the wall and the tube will lose its strength. Several types of tubing have been used successfully for race cars. 1010, 1015 and 1018 steel in welded, seamless and D.O.M. are the most popular and weld easily. Use 1 ¾-inch by .095 inch wall tubing in the main cage. Notching tubing uses the same methods as for pipe.

Cage Kits

If you are going to farm out much of the bending, you will be way ahead to purchase a kit from one of the several manufacturers. More and more people are going this route because by the time you buy the tubing, rent a bender, and find some way to notch it, you have more than paid for a kit. Most manufacturers have a bottom line kit called a "Street Stock" or "Budget" kit. These are marginal at best. The second line of kit is usually a better buy as you would have to add bars to the cheaper kit anyway. Make sure your kit has most of the notching done for you or keep shopping.

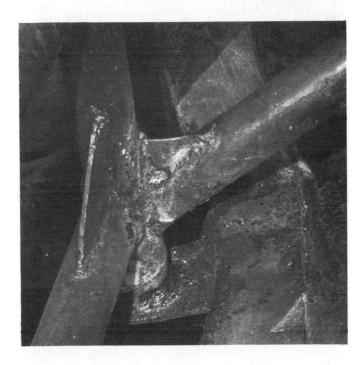

Don't hang the leg of the cage outside of the frame unless it is well supported.

Notice how nicely the cage fits the sheetmetal here. Also note that the cage is well triangulated and there are no unneeded bars here. You can see at the base of the left rear tube how a plate was used to tie the tube into the unitized body.

The plate on the main hoop (arrow) allows you to make the roof hoop wider.

Chevelle frame was boxed in to mount the cage.

No matter which material you choose, you should use gussets at all joints. Many track rules require them and the small amount of weight they add is a fair trade off for the additional strength. Unless you have access to a shear that will handle 1/8 inch thick material, buy the commercial gussets. By the time you cut them out with a torch and grind the slag, it will cost you more to make your own than to buy them.

A word about welding. If you can't weld, don't use your cage for practice. Get someone who knows what he's doing to weld the cage. You can practice welding on the bumpers and other areas that are not critical for safety. If a shock mount falls off, no one gets hurt, but if you get a cold weld on a sidebar upright it could mean a stretch in the crash house.

If you are reading this book, then you don't race for a living. This means that you have to get up Monday morning to go to work. The point I am trying to make is "Do not make compromises in your safety equipment, including your cage." Racing is dangerous enough with good equipment, so put some thought into the design of your cage. Don't put a bar somewhere just because it fits there. Every bar in a cage is there for a purpose. When designing a cage, keep in mind that a triangle is one of the strongest shapes known to man. When you look at a Grand National cage, you will see that it is a series of triangles. This puts these bars under compression (pushing the ends toward each other) or tension (pulling the ends away from each other.) The diagrams show several ways of building a main hoop that is structurally sound for both a rollover and a side crash. Side bars on at least the driver's side should be bent. Bent bars give more crush area in a bad crash. If you have a chassis that is quite a bit narrower than the main hoop, make sure that it is tied into the frame so that the load on the bars which hold the main hoop to the frame does not have to support the weight of the car if it gets upside down (see diagram).

Building A Cage

In building a cage, you start with the main hoop. Rather than just making two bends at the top of the hoop, it will make overall construction simpler if the main hoop has two bends on each side. This allows the roof hoop to hit further outward inside the roof than if there were just one bend in the main hoop. The wider the roof hoop, the more head protection. The roof hoop can now be connected to the main hoop using quarter-inch plate. All of the inner panels in the roof should be cut out before you start fitting the cage. Make sure the roof hoop fits up tight over the windshield so it doesn't obstruct vision on banked tracks.

The bars at the windshield post can be done in two ways. The first is to use one bent tube or the windshield post bar and front door bar. The other method is to use a long straight bar for the windshield bar and put in a second vertical bar for the front door bar. Whichever method you use, make sure that the front door bar is parallel to the vertical bar in the main hoop. This makes it much easier to fit the side bars since the length of each will be the same.

You can now tack in the dash bar and additional bars in the main hoop. Try to fit the side bars as close to the sheet metal as possible. This saves a lot of sheet metal work. To save work, some people use straight bars on the right side of the car. In the long run, unless your track rules allow rub rails outside the sheet metal, you will probably spend more time doing body work than you would have spent bending the bars in the first place. There are some designs that have one of the side bars run from just behind the front wheel to just in front of the rear wheel to protect the sheet metal.

It is easiest to tack the cage together in the car, than to either remove the body or the roof. Many people remove the cage from the car so that it can be welded on the floor or bench. This way it can be rolled around the floor and makes the welding easier.

If the cage is fitted and welded in the car, be sure the inaccessible tops of cage tubes get welded somehow. Either cut the window posts with a hack saw and lift up the car's roof to provide access, or else cut a hole in the roof with a torch to provide access.

Mounting The Cage

A roll cage is only as strong as it's mounting points to the frame. Simply welding it fast in any old way will not do. The stronger the frame, the better the mount will be. If the frame is weak, then it must be beefed up at the mounting points.

This windshield post used one piece of tubing to follow the angle of the post, and then a second vertical to support it. The vertical makes it easy to weld in the door bars.

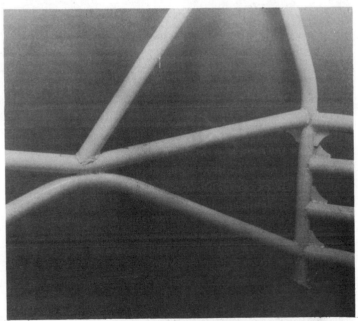

The design of this cage gives support to the door bars.

The padding on the sidebars saves a lot of bruises on the driver's knees.

This bar from the roof hoop to the top door bar is not the world's worst idea. It protects the driver's head if someone's bumper tries to climb in the cockpit with him.

A frame that is a box section is the easiest to work with. In this case, you can cut a hole in the top of the frame with the correct size hole saw to fit the tubing. This allows the vertical legs of the cage to be supported by both the top and bottom of the frame section even though only the top is welded.

A frame that is a channel section should be made into a box. This can be done by cutting the top section of the channel and replacing in it a section of rectangular tubing. This seems to be a little easier than boxing in the channel with 1/8 inch thick plate as many people do, even though it is a little heavier in weight.

A method used on the GM channel frames that have the top of the channel at an angle, is to take a short section of heavy angle iron and weld it to the side rail. This gives a flat place to mount the cage.

Cage In A Unitized Body

If you must mount the cage to the floor pan because the track rules will not let you put a frame under a unitized body, then try to spread the mounting point to as large an area as possible. Use 3/16 or 1/4-inch plate of about one square foot in dimension. Weld or bolt this to the floor pan. If it is bolted, use large flat washers under the pan and at least 3/8-inch bolts. The feet of the cage can be welded to these plates.

This welded-on outrigger gives a flat, sturdy place to weld the cage.

Summary

Pipe cages are cheap and safe if well designed, but they are heavy. (35 to 65 percent heavier than .095-inch tubing). A home built tube cage is difficult unless you have the right equipment. Kits are a good way to go if you are short of time and/or equipment. The design and welding itself are very important in constructing a safe cage. If you are going to design your own cage, look over those in well built cars for ideas.

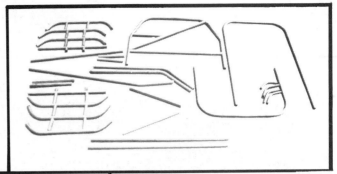

The Speedway Engineering sportsman roll cage kit.

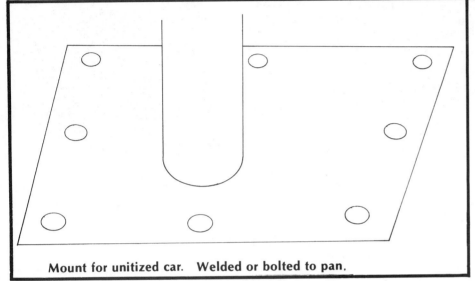

Mount for unitized car. Welded or bolted to pan.

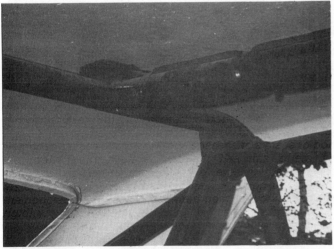

A roof bar over the driver's head is not a bad idea, safety wise. Since no cages fit the roof line exactly, this builder welded in a piece of channel for a spacer and a place to bolt the roof. The bar placed straight like this offers no structural rigidity to the cage. However, if the bar was installed at a diagonal from the main hoop on the right to the hoop above the left windshield post, it would help the rigidity.

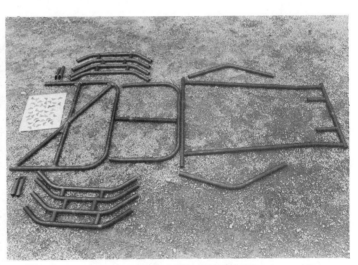

The Howe Racing roll cage kit.

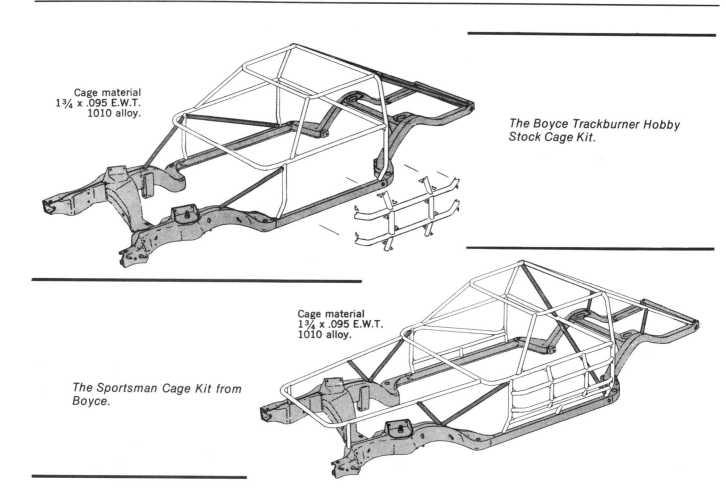

Cage material 1¾ x .095 E.W.T. 1010 alloy.

The Boyce Trackburner Hobby Stock Cage Kit.

Cage material 1¾ x .095 E.W.T. 1010 alloy.

The Sportsman Cage Kit from Boyce.

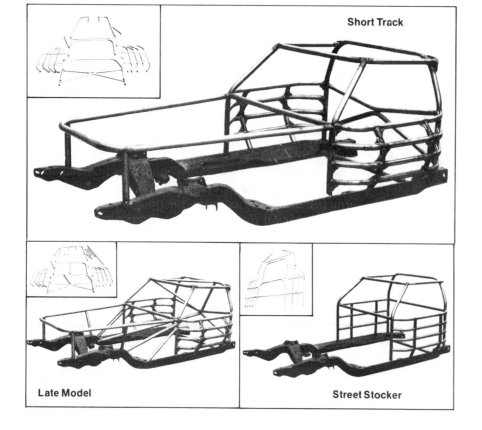

Short Track

Late Model

Street Stocker

Cage kits from Canadian Stock Car Products.

Chapter Three
Supports in Addition to Main Cage

Most tracks allow you to install additional supports both in front of and behind the cage. If your track allows these supports, by all means weld them in your car. If your car is not constructed as a space frame you will have chassis flex. When this happens, you are using the frame as a suspension component. Because the frame is not spring steel, this will cause the car to be erratic in its handling characteristics. It will push in one corner and be loose in the next. You will also find that you will be running stiffer springs with a flexing chassis than those cars that have better support of the suspension mounting points.

Rear Supports

The supports in the rear will depend upon whether you use leaf springs or coils. If you run leafs, you will want a support running back to the rear shackle mounting point. With coils, it is not necessary to support this far back for suspension strength, but you may wish to for support in the event of contact in the rear with a moving or stationary object. As we mentioned in cage construction, try to triangulate everything. Also, on any long span of tubing (four feet or more) put in intermediate supports.

In order to save weight (and because the rules do not require it), many builders use smaller diameter tube for rear supports. If weight is not a consideration, this is a good place to use up that pipe you got so cheaply. It is better to use it here than in the front. This keeps the weight bias to the rear.

If you used plates to connect the roof hoop to the main hoop of the cage, this is where you want to connect the supports which run to the rear of the chassis. If you weld these supports to the main hoop, in the event of a bad rear end crash, there is the possibility that you will weaken the main hoop. The weakest point on the rear chassis is the section of frame where it kicks up over the rear axle. In the event of a crash in the rear of your car, it will push this section forward and bind up the suspension, be it coils or leafs (see figure 1). To avoid this, you should run a piece of tubing from the frame kick-up forward to the main hoop vertical in the area of the side bars (see figure 2). Run another bar from the kick-up to the center of the main hoop and you will have a strong, well-triangulated unit (see figure 3).

In many cases, when you use a stock chassis, there is a problem with the rear end housing hitting the frame during normal suspension travel. If you cut an arch out of the frame for clearance, make sure that you do something to replace the strength that you have removed by making the cross section of the frame smaller. This can be done by

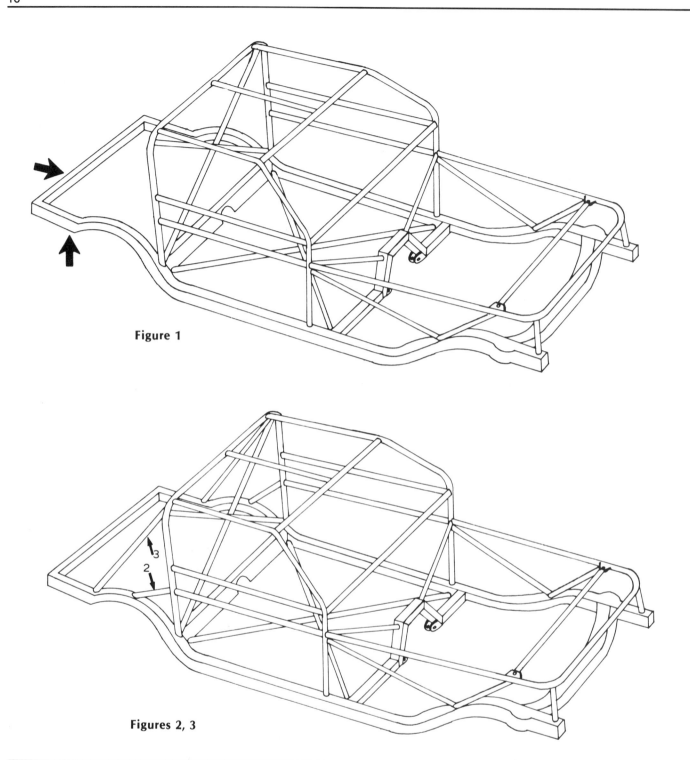

Figure 1

Figures 2, 3

plating the sides of the frame at this point with .125" plate.

If your stock chassis in the rear is a channel section, such as the Chevelle type of frame, you should either box in or strengthen it with tubing or pipe. If you don't, every little rear end tap will mean a bent frame and wrinkled sheet metal. If you are a beginner, there is a strong possibility that you are going to get rooted a few times by the faster cars. Maybe you won't need the extra support after you become a hot dog, but the extra weight won't hurt you for a while, and it will save a lot of frame straightening.

Front Supports

For good handling, you must support the suspension points, (especially the right side) because of the heavy loading they receive. Your second consideration is to strengthen the frame from a front end crash. Any area of

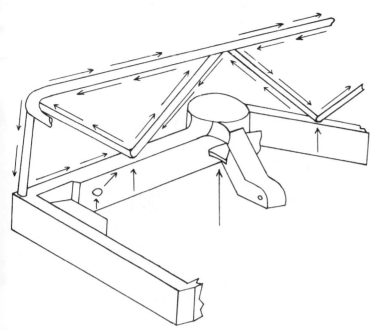

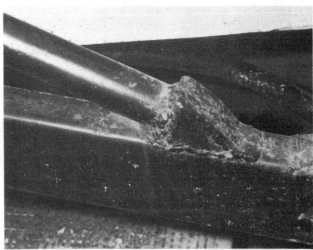

A torque box requires more work, but it spreads the load from the tube on the frame.

Above, the arrows indicate how a load fed in through the springs acting against the frame is fed into a properly triangulated chassis front bay. Notice that all of the tubes are completely straight with no bends or radii. Any bend will serve as a weak spot, allowing the structure to flex at that bend. Below, the arrows show how an untriangulated structure is flexed through bending loads imposed on it.

The rear section of the frame has an X. This adds considerable strength. The fuel filter has been wrapped in foam rubber to keep the pounding from breaking it.

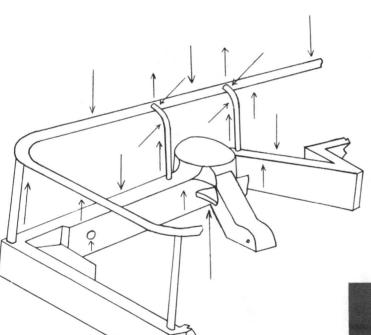

Two bars on top of each other is poor planning.

the frame with a bend in it will be the first place to kink even in a minor crash if it is not braced. Do some planning before you start to weld. Try to accomplish both objectives with the same bars whenever possible as this will save weight and money.

One bar that will serve both purposes is the bay bar. It runs from the center of the main hoop to the frame just behind the upper a-arm on the right side. This is quite a long span so it must be supported at one or two places. The bay bar itself should be at least 1¾ inch O.D. x .095 inch wall, while its intermediate supports can be lighter.

Most cars run a front hoop of some type. This serves to protect the radiator, mount the shocks, gives a base to mount the fenders and hood as well as help you make the front into a space frame. The hoop itself does nothing to stop frame flex, but with a few short pieces to triangulate each side, it becomes quite strong and will resist some twist and buckling.

Make sure the front hoop is mounted to the main cage at the side bars and not just to the dash bar. Most builders put a very small bend in the front hoop, about two feet in front of the point where the hoop attaches to the cage. In the event of a very bad front end crash, the hoop will bend at this point rather than break at its mounting point and possibly come back into the cockpit area.

Bumpers

Most hobby classes require that you run stock bumpers, but do allow you to strengthen them from behind. Some tracks allow tubing to be added above the stock bumpers for additional protection of the grille and rear sheet metal. If your track permits this extra protection, take advantage of the rules.

In designing your bumpers, keep in mind that thtal. If your track permits this extra protection, take advantage of the rules.

This bumper is designed to give before it tears up the frame.

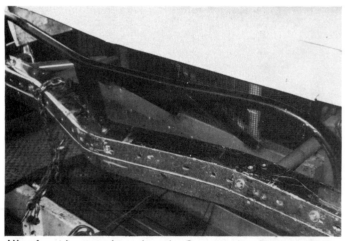
Nice front bumper based on the Camaro aluminum bumper.

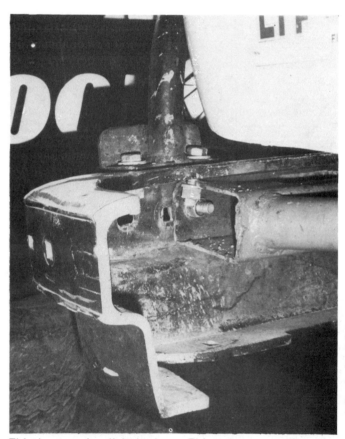
This bumper is all bolted on. This takes more time, but seems to save the chassis.

Note how the radiator protection bars are mounted so they will not tear up the front hoop in a crash. This type of screen does a good job protecting from rocks and stray tie rod ends.

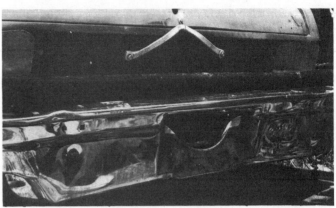

Stock bumper is only bolted to the heavy tube behind it. Note trunk holddown.

In designing your bumpers, keep in mind that they are going to get bent no matter how strong you build them. If you make the bumpers too strong, the frame will get bent before the bumper. Your bumpers should be built so that the bumper will absorb the crash rather than have it transferred to the chassis. Try to allow for a crush area so that you can control where the bending will take place and not have it affect the handling. Don't tie the front bumper into the radiator supports, front hoop or the hoop uprights. If the radiator support gets bent, in most cases, you will have a leaking radiator

Nerf and Rub Rails

Nerf and rub rails are not allowed at many asphalt tracks, but are sometimes legal on dirt tracks because of the wider tires. They do save a lot of sheet metal work if permitted. Make sure that there are no open ends which can cut your tires if the bars are accidentally bent into them. One way to prevent this is to cap the ends with the cup type of freeze plugs. These are available in just about any size needed for tubing.

One thing that might be allowed on a track that requires stock appearing cars, is the aluminum side molding from school busses. You can bolt it to the sheet metal or make short brackets to the interior tubing. If you mount it at hub height, it stops a lot of the donut marks from tires rubbing, even though it will not stop damage from a hard crash. When painted the same color as the sheet metal, it is hardly noticeable.

If you use rub rails of any type, be careful how they are mounted. If your rub rails get bent, you don't want the supports for them to be mounted so that the cage or door bars suffer a rip or tear. I have seen rub rails mounted directly to the cage so that if the rub supports broke loose from the cage, they might spear the driver in his seat. Rub rails are just as the name implies—to protect against sheet metal damage in a minor brush with another car. In a T-Bone crash, they should be designed so that they just fold up against the side bars of the cage and allow the shock to be distributed over a large area.

Summary

Look at other cars to see how they do it. Try to figure *why* each bar is in the car. Compare cars so you can learn how one person used one bar to accomplish something that it took another person two or three bars to do the same thing with. Look at cars after they are crashed. Did the design work so there was a minimum of chassis damage or did the bumpers only get scratched while the

Super way to mount bumpers. The rectangular tube will bend first. This absorbs a lot of the force so it is not transmitted to the frame.

chassis will have to go on a frame machine before it can run again? Look at some of the poorer handling cars. Are they missing a few bars that the better handling cars have? THINK!!! Don't weld in a bar just because it happens to fit there!

Open tube like this will do a job on that new tire.

The rub rail comes out almost to the outer edge of the tires.

Very nice bumper, nerf bars and rub rails.

Welding the threaded end of pipe is ok in non critical places like this rub rail, but is a very poor practice anywhere.

Capped off rub rail.

This Camaro has the flexible type plastic front end which seems to work well.

The hoop on the front bumper should have some support or it may bend back into the radiator with a light crash.

Neat way to mount rub rails by bolting.

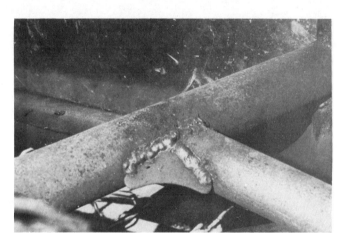

This rub rail support is not well planned. When two tubes meet in the middle of a span, it doesn't take much of a shot to bend everything.

A freeze-out plug tacked to the nerf bars may save you a tire. The rough end of pipe or tubing cuts like a razor when it comes in contact with a moving tire.

Chapter Four
Frame Chassis vs Unitized Body

Advantage of a Frame Chassis

A stock car is not a race car unless it has a full frame chassis of some kind under it. A unitized body is fine for the "Stoplight Speedway" or Indy type racer, but theirs is not basically a contact sport. A beginner should choose a car with a full frame under it for the following reasons:

1. Where do you hang the cage? If you weld the cage to the floor pan as many novice racers have done, you had best mount the seat to the cage so when the cage gets knocked out of the car, you can go with the cage.
2. All race cars get crashed sooner or later. For beginners, it is usually sooner. With a full frame chassis, when you bend the frame in a crash, you go behind the bend, cut off the bent part and weld in another section of frame. With a unitized car, you go to the local body repair shop and have them put the car on their frame machine while you go to the bank to float a loan to pay for it. Take your time because unless the guy doing the work is your buddy, he probably has better things to do than mess around with an old race car.
3. To cure the above two problems, many racers tie the front stub to the rear stub of a unitized car with rectangular tubing. This can be 2 x 4 or 2 x 3-inch tubing with .125-inch wall. A perimeter frame is made with the tubing under the floor pan to leave a stock appearance. It is necessary to cut out some of the tin under the car so this homemade frame fits up out of sight. Access holes must be cut in the floor so that joints can be welded. All this should bring the car in at about 3600 pounds so it is not only a lot of work, but is heavier than a stock chassis.

Chassis Under Unitized

A compromise that many tracks have come up with in their hobby classes is to allow the racer to put a chassis from one car under the sheetmetal from another. An example is, a Chevelle frame shortened seven inches placed under a Camaro body. This is not as difficult to do as it sounds if you have torches and a welder. It basically amounts to taking the skin off two cars and welding or bolting the desired body on. If stock floor pans are required, use the pan from the chassis after removing the rubber body mounting bushings and mounting the floor pan solid to the frame. Sheet metal can be used to fill in any gaps. If stock floor pans are not required, cut the pans out and fabricate your own from 20-gauge sheet metal using one-inch angle iron for bracing.

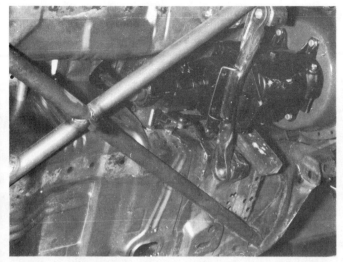

These three photos show another way to tie together a unitized body. Plates were welded each major force input area of the chassis, then sheet tubing was used to triangulate these points. After seeing all the time, work and money it takes to make a unitized car right, doesn't it make sense to use a conventional frame chassis?

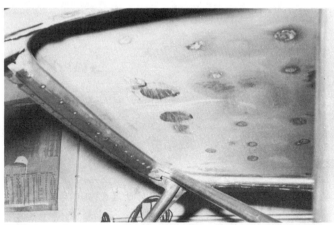

Note how important it is to get the cage bars close to the sheetmetal. The sheetmetal is then tacked to the cage to make the total structure stronger. Note how the roof putty "junk" has been cleaned out.

Stock floor pans are not only heavy metal, but like much of the sheet metal in a car, are covered with a tar-like material used to deaden sound. There is no fast way to remove this material, but if there are a couple of kids in the neighborhood who want to help with the race car, give them each a putty knife and turn them loose on it after you have heated the material with a torch. The same goes for the junk inside the roof.

At some tracks, they let you fabricate your frame as long as you use a stock front stub. This is done with 2 x 3 or 2 x 4 inch rectangular tubing. I don't recommend that you do this for a first car, but if you insist, lay everything out on paper first. You don't have to be a draftsman as you can use graph paper.

In making your own design of frame, there are a couple of things to avoid. Make sure that there is enough clearance over the rear axle housings. Make sure there will be sufficient tire clearance as the body moves sideways as does the sidewall of the tire. This is particularly important in a car without a Panhard bar, or when rules require a tire with a high sidewall. Where the frame comes down in the back, it should be about the height of the middle of the bumper for strength. Tack everything together first to make sure it all fits before welding it solid. It is a lot easier to cut a tacked piece out than something that is completely welded (and all of us do make mistakes!).

If you are using a stock chassis and wish to get the car lower, you can cut V-shaped sections in the top of the frame just behind the front stub and just in front of where the frame kicks up in the rear. If the open end of the V is about ½ inch, it will drop the car about one and a half in-

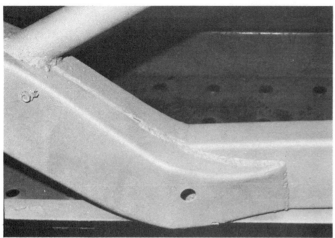

Above and below, a very neat mating of a Camaro stub with 2x4 tubing chassis.

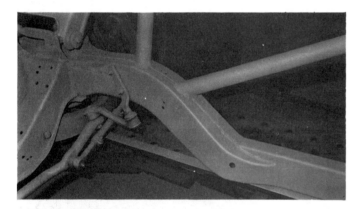

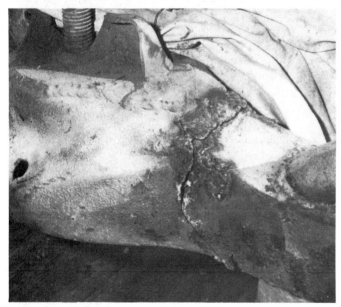

This crack has been welded once. About time for another front stub.

ches. Rub a little mud on the weld so it looks likes the rest of the chassis and no one will ever see it. Try to make any chassis modifications at a factory weld if you are fudging on the rules. This makes them harder to spot. If need be, knock in the modified section with a hammer and fill with body putty. Add a little paint and nothing shows.

Interior Sheetmetal

Even if your car is stock inside, once you remove all the upholstery and rugs, there will be holes to cover. The simplest way is to use pop rivets. Use the steel rivets rather than the aluminum. For your use, the aluminum ones usually loosen up from the pounding a race car takes. If you break a drill bit, (and you will) don't throw it away. Resharpen it. When doing sheet metal work, these shorter bits last longer than the new ones. Don't skimp on rivets, or you will have pieces of metal floating around while you are racing.

You will have had to cut out some of the floor pan in order to weld the feet of the cage, so these areas will have to be covered. One quick and simple method is to cut a circle from a piece of sheet metal which is the size of the tubing. Slit one side into the circle. The piece will now slip over the tube and make a neat installation. If you had thought of it before the feet of the cage were welded in, you could have cut out the hole and slid the sheet up the leg while you were welding.

If you are redoing all of the interior metal, do a little planning. Make those sections removable that will aid in maintenance—such as over the transmission. You can tack weld nuts to the angle iron bracing so it is easy to remove the bolts holding these removable sections. In stripping out the interior of your car, you probably will find quite a few nuts attached to clips. These quarter-inch nuts can be used for the removable sections also. Ford makes a bolt and nut combination for putting sideview mirrors on sheetmetal that locks the nut on the metal, allowing the screw to come out and leaving the nut in place. These could work for your purpose.

Cutting sheetmetal patches to fit around tube as described in the text.

This is a Camaro front stub mated to a Chevelle frame. It's cheap, but a lot of work.

Large pieces of cardboard should be used for templates to cut the pieces of metal. Body shops usually have a lot of large pieces from boxes which body parts are shipped in.

Spindles and Ball Joints

Unless your track allows extremely wide tires, you can get by with the stock spindles and ball joints in most cases. If you try to go to a larger spindle, make sure that your ball joints will fit the spindles. If the taper in the spindle is too small, you can get it reamed out to the correct size by a shop that specializes in race car work, but most normal machine shops will not have the correct reamer. It is better to use the correct ball joint for the spindle and have the A-arm bored out to fit the ball joint. Unless you feel that the larger spindles are an absolute necessity, leave well enough alone. It can turn into a real zoo if you don't know what you are doing.

Some racers wish to convert to disc brakes on the front of the car. If you feel that you need these rather than your drum brakes, the conversion will be easier if you convert everything from the master cylinder forward in the brake system. This means that you should use all the plumbing from the disc brake car as well as the master cylinder and spindles. This way you don't have to know much about brake engineering.

When track rules require safety hubs on the right front, try to find some used parts you can buy from a sportsman or modified car. Many of these cars update their equipment from time to time and discard used suspension parts. If you can't find something used, then it becomes a pretty expensive proposition. MRE has about the cheapest setup because it uses a Chevy rotor and caliper to go with their spindle adapter and hub.

If you use the stock cast iron hub, check for cracks after each night of racing. If you have a machinist that is a very good friend, you could get him to make you a hub out of steel that will take larger bearings. If you had to pay for someone to do this, the cost would be high.

Reworking A-Arms

When you drop the chassis down low enough to gain a decent camber curve, you may find that the upper ball joint is at a very sharp angle. To check to see if there is sufficient movement in the ball joint, remove the front springs and lower the chassis with the front wheels mounted until the frame sits on the ground. If the upper ball joint keeps the frame from reaching the ground, then the upper arm should be modified to give the ball joint more travel. Simply cut the vertical bracing on the arm about three inches in from the ball joint. Bend the horizontal section of the arm down some to make the ball joint flatter. Then weld the arm back up with some 1/8-inch plate for added strength where it was cut out. It will not be necessary to do this to most cars as they have enough movement in the ball joint.

When using wedge bolts (weight jackers), it will be necessary to cut clearance on the upper arm so that the arm does not hit the bolt under normal suspension travel. If you wish to strengthen the arm after cutting, lay two

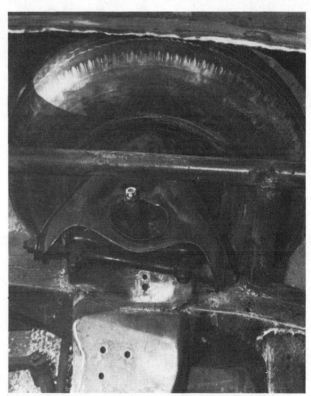

This upper arm still has to be cut from wedge bolt clearance.

small pieces of pipe or tubing on the arm and weld. It may be necessary to bend the tubing slightly in order to follow the contour of the arm. The lower arm can be beefed up in the same way or by boxing in the arm with 1/8-inch plate. When welding on control arms, remember that the inner bushings are rubber and too much heat will destroy them. It is best to replace the inner bushings with new hard rubber ones—or better yet—replace them with soft metallic ones.

While discussing these bushings, you will find advertised, hard rubber or steel bushings to replace these stock bushings. I don't feel that it is necessary to spend the extra money for a hobby car. Naturally there is some compliance in the rubber, but most hobby class drivers can't tell the difference anyway. Of course, if the bushings are worn out so they are sloppy, replace them with stock parts. One advantage that the rubber mounts gives you is the fact that in a minor crash, there is a little give so that nothing gets bent. With solid mounts, something has to give. If you have beefed up the inner mounts for the upper arm, you can take a pretty good shot on the wheel and not bend the arms.

Shock Absorbers

The shock absorbers which come with a passenger car are fine for the street, but if used on a race car you will find yourself bouncing up and down all the way around the track. The valving in passenger car shocks is completely different than that found in race car shocks. The racing shocks you choose should not be so stiff that they restrict body roll, but should be stiff enough that they can control the spring from bouncing the car up and down when it hits a bump or takes a set in the corner.

Even in a "strictly street" class you can replace the stock shocks with Carrera racing shocks. These are stock replacement shocks, but have the proper valving for racing requirements. They bolt right into the original mounts.

When using wedge bolts, it will be necessary to fabricate your own mounts for the shocks. In this case you will use regular racing shocks such as Monroe, Gabriel or Carrera. These come in two basic types—tie rod end and rubber bushing end. Hobby racers opt for the rubber bushing type because they are a few dollars cheaper and the tie rod end shocks require a tapered hole to mount. Use a large washer on the rubber mount shock so that the bolt can't pull through the rubber or the rubber can't pull out of the shock.

Racing shocks come in three lengths which allow from six to ten inches of shock travel. You can only use half of the travel as the shock should be mounted about in a half-

A Carrera racing shock to replace standard road shock.

The side of the spring pocket was cut out so the shock could be mounted as close to the ball joint as possible.

This shock was spaced out from the bracket so the shock body would not hit the tube.

When you use a rubber bushing end shock absorber, be sure to use a large flat washer on both sides of the rubber bushing so the nut will not pull through.

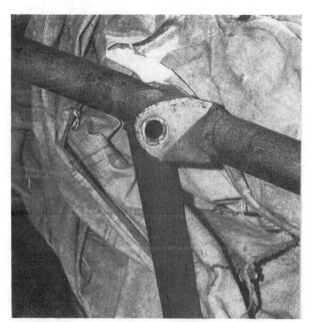

The shock mount is strongest when it is attached at the intersection of at least two tubes. This gives added support to the input loadings fed in by the shocks.

way position. The shortest travel is used in the front and the medium or longest in the rear.

Make sure that there is enough travel so the shock does not bottom out. There are rubber bumpers on the stem of the shock. These can be used to see how much shock travel you are getting at each corner of the car. If this rubber is pushed all the way up after you run the car, the shock should be remounted so it has more travel. If a shock bottoms out, it will cause the car to break away on the end the shock is mounted. For example, if the right front shock bottoms, the car will push no matter what you do to the suspension because the shock then becomes a spring of infinite stiffness.

Sometimes you want to try a softer shock, but don't want to spend the money just for an experiment. If you remount the shock to a slightly greater angle than the original mount, you will have the action of a softer shock. The more vertical a shock, the stiffer it is. The more angle the shock has, the softer it acts. But don't get too carried away with this mounting angle—especially on the left rear. If too much angle is on the left rear shock, the oil in it tends to climb up the wall of the shock and causes aeration.

Leaf Spring Wrap-Up Control

There is another type of shock that is made by Carrera which is used to control spring wrap-up in a leaf spring rear. It mounts from a bracket on the rear end housing to a bracket on the chassis or cage. Because of the specifically tailored dampening action, this shock should not be used for any application other than that for which it was designed. This type of shock is also available from Baldwin Engineering.

Seat and Belts

If you must use a stock seat, weld or bolt the seat adjusting slide so that the seat will not move around in case of an impact. Also, run some bracing from the roll cage to the seat back if the back is not already against the cage. Run two pieces of strap from the floor in front of the seat, over the seat to the floor behind the seat. Bolt this strap to the floor. Use very large washers under the floor and make the straps very tight.

A regular racing seat is easy to mount and much more comfortable to use. Both the fiberglass and aluminum seats fit around you and help to hold the driver so he

Use the right length bolts. It's less work and looks more professional.

The angle on this shock is such that it will have a normal rate in comparison to the other two adjustments available for it. Note how the brake line is tie-wrapped to the axle housing. This brake system is well-plumbed.

Use a large flat washer on rubber bushing shocks or the nut will pull through.

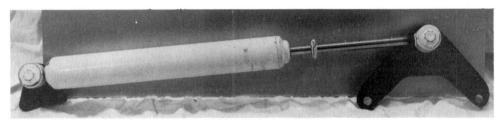

A Carrera "Axle Damper" shock for leaf spring cars. Runs horizontally from bracket on the rear housing to the case or frame to eliminate spring wrap up.

doesn't have to hang on to the steering wheel to stay in the car. Your belts will hold you in better since there is no give in the seat as there is in a passenger car seat. A framework should be made for the seat out of light tubing or 1/8 by 3-inch plate. This gives you something to bolt the seat to. You shoud have a minimum of two bolts in the seat bottom and two in the back. A piece of plate should be used inside the seat to tie each pair of bolts together and sandwich the seat itself between two pieces of metal.

The belts themselves are what hold you in the car and should have their mounts welded to the chassis or cage. The lap belt should go across the pelvic area and not the stomach. The belt mounts should be about two inches forward of the line of the drivers back. Some tracks require a chain to connect the belt mounts and run under the chassis. If this is the case, be sure that the chain does not hang down so it can catch on something.

The shoulder harness is designed to not only keep the driver from going foreward, but also to keep his rear end planted in the seat in case of a rollover. This should be mounted to the roll cage slightly below the level of the

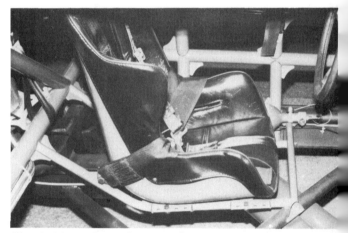

Check this neat seat framework.

driver's shoulders. Most people don't use an anti submarine strap unless the seat lays at quite an angle. think there is a subconscious fear of talking with a high pitched voice after a crash. But seriously, they should b used to keep the seat belts in place in case of a hard impact.

Keep your belts in good shape. If they get cut or worn out, replace them. Have the belts as tight as possible when racing, because if they are loose, they won't do much good. Adjust the lap belt tight first, then pull down on the straps of the shoulder harness to tighten it.

There should be a cotter pin in this type of mount. A flip could unhook this belt mount.

A good, substantial seat belt bracket welded to the frame.

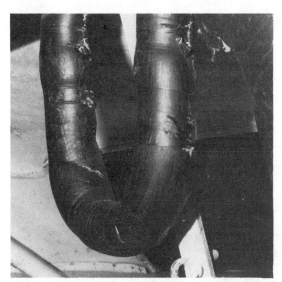

Better put some more rubber on this solid headrest or you may get a headache.

A headrest this size is hardly worth the trouble. You'll probably miss it in the event of a crash.

Every bar in reach of the driver's body should be well protected by ample padding. Note how everything is within close reach of the driver.

Almost any type of universal is better than the stock "rag joint."

The steering has been lengthened by welding in a section of ¾-inch I.D. pipe. When you weld anything in the steering system, make sure the welds are perfect.

This aircraft type universal was welded and bolted. Not a bad idea if you are unsure of your weld.

White interiors are hard to keep clean, but give you much more light when working in the dark. Not how section over transmission bolts in for easy access.

Well mounted master cylinder so it will not flex. Note how to make sheetmetal go around a curve.

A simple way of mounting body panels with 1/8 by one inch strap.

Chapter Five
Engine, Cooling and Electrics

Engines

As stated before, don't get carried away with your engines. Chances are, you won't be able to use all the power for a while anyway. Build yourself a fairly stock, dependable engine. Stay away from high compression and high RPMs. Both make the motor undependable unless a lot of expensive parts are added for insurance. Talk to several engine builders about their recommendations and try to find one that has ideas similar to your own.

Most classes of hobby cars allow you to use a flat tappet camshaft of your choice. If you use a short duration cam, it gives you low end power without turning high RPMs. This gets you off the corner and unless your track has extremely long straights, you won't run out of power before you have to back off for the next corner. TRW puts out a line of high performance camshafts that do a good job and are quite reasonable in price.

Since most oil pumps drive off the bottom of the distributor, do not beef up the oil pump unless you use a bronze gear on the distributor. The added strain causes the distributor gear to wear which in turn chews up the gear on the cam—a very expensive problem. If you find your timing changing, pull the distributor and check the gear. It is not necessary to run a great deal of oil pressure anyway.

If your track allows you to run a four-barrel carburetor, it is not necessary to go out and buy the biggest carb made. If you over-carburete an engine, it will make it sluggish off the corner and will take the engine half the straight to catch up. The Grand National cars go quite well with relatively small carburetors.

There are linkage kits on the market to convert from vacuum to mechanical secondaries if you can't afford a Holley racing carburetor.

If you buy a used engine from someone, take it apart before you run it so you can inspect everything and at least do a valve job. It has been found that a valve job will pick up the power on a used engine quite a bit. The disassembly and inspection, if nothing else, will give you piece of mind that the engine isn't ready to unload the next time it is run. Check the bearings and rings for wear and end-gap. Check the pistons for cracks. Check the oil pump for wear and nicks and the cam for lobes starting to go. Look for broken valve springs and cracked rocker arms. See if there are any hone marks left on the cylinder walls.

Don't be afraid to put an engine together yourself if money is tight. Get a shop manual from the dealer for the motor you're working with, have an engine man do the machine work and then put it together yourself. Take your time, be careful and keep everything clean. If it's a basically stock engine and you can follow directions, you'll be all right. There is nothing like the feeling of putting together your first engine and having it run.

Either replace the rubber engine mounts with steel mounts or weld on some sort of restraining tabs so that the engine doesn't fly out of the car in the first crash.

Pans

If you find that you are having fluctuations in oil pressure in the corners, it is probably being caused by a lack of oil to the pickup. An engine suspends quite a bit of oil above the crankcase at higher than stock RPMs. You can try running one or two quarts more than stock. This may help your problem of pressure fluctuation, but may also cause the engine to blow oil out the breathers. Oil tends to climb the right side of the pan in a left hand turn. This takes the oil away from the pickup. To compound the problem, the oil goes to the front of the pan under deceleration and to the back of the pan under acceleration. This is why most of the high buck engine builders now only build their engines for dry sump systems which you can neither afford nor legally run.

You can modify your pan and pickup to make it more efficient, though, by putting an extension on the side of the pan and extending the pickup into the extension. Bolt a stock pan and oil pump on a bare block. Bolt a starter to the block or bellhousing if the starter mounts on the right side of the block. Cut a section out of the sump (deep) area of the side of the pan. Fabricate a box out of sheetmetal to be welded to the pan. Make sure this allows you access to the starter bolts so the starter can be removed without taking the pan down. Before you weld the box to the pan, add tubing so that the pickup extends into the box and is 3/8-inch from the bottom of the pan. Tack the box to the pan. Take the starter off. Take the pan off to make sure that you can get it off and on. In all the previous steps, you only needed a few bolts holding the

If the engine has been moved in the chassis you may have to do some design work so you can work the clutch. This linkage should have a triangular mount to the frame at the relay rod pivot to prevent any flexing. The relay rod could use a support at its other end is well.

A solid engine mount for a six cylinder.

If you use stock motor mounts (rubber), you had better chain down the engine.

pan to locate it on the block. Before welding the box to the pan, bolt down the pan with all the pan bolts so that it will not distort from the heat. When done, fill the pan with water to check for leaks and check the pickup tube where it is welded to make sure it can't suck air. The pickup tube should be brazed to the pump base so it cannot fall off. If the pickup tube is long, you should fabricate a support to be bolted to one of the main cap bolts.

This will allow you to carry more oil and make it available to the engine at all times. The oil to the rocker arms should be restricted somewhat in a racing engine. If not, you will be blowing oil out of the valve cover breathers. This can be done in many engines simply by inserting the cam bearings so that only part of each hole lines up with the holes in the block.

Becoming popular is a system to furnish oil to the engine under demand. This is done by using a pressure tank (about three or four quarts capacity) that has a line running to the main oil galley. When the engine has oil pressure, this forces oil into the tank at whatever pressure is in the engine. If the oil pressure drops in the engine, the pressure of the tank forces oil into the engine until the pump picks up the pressure. This system is available commercially or you can make your own. If you make your own, be sure that the tank is pressure tested to a much greater pressure than it will be required to hold. The lines should be large diameter, aircraft quality, high pressure hose. This system is good for pre-oiling an engine after it has not been run for a while. For this reason, install a shutoff in the line so that you can hold oil in the tank under pressure.

A box was welded to the pan under the starter as described in the text.

This is a homemade pressure tank for preoiling a Mopar as described in the text. Electrics are mounted on the dash for less heat and easy access.

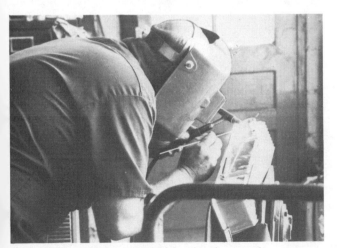

Installing fitting for breather in aluminum valve cover.

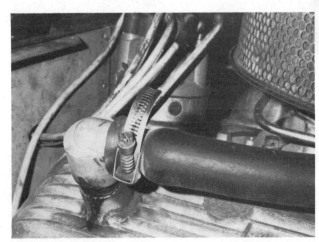

A fitting was heliarced into these aluminum valve covers and the hose runs forward to the breathers.

Radiators

A cross flow radiator is better than the older vertical tube, but many newer passenger car radiators are only two tubes thick and usually won't cool a race car. Check out the junk yards for a thicker radiator that will fit your application. It doesn't have to be an exact fit as you should fabricate braces for your race car radiator better than what comes stock on the car. Use at least a 15-pound pressure cap. The higher the pressure, the hotter water can be before it will boil in the system.

While atending to the cooling system, you should look at thermostats and pulleys. If you can slow down the water pump by means of a larger pulley on the pump or a smaller pulley on the crankshaft, it will slow the flow of water through the radiator which in turn lets the engine run cooler. You also can slow this flow by using a restrictor in place of the thermostat. The restrictor is made by cutting a piece of sheet metal in the shape of a disc that fits the thermostat housing and drilling a hole in it for the passage of water. Since these are so easy to make, make several with different size holes drilled in them so you can experiment to get the temperature where you want it.

Since most tracks require a catch can of some type to avoid water on the track from overflowing radiators, the simplest thing to use is a stock type recovery system. If you want to get fancy, you can use a surge tank setup from a Corvette, or make your own surge tank. Make sure the filler for this is at the highest point so you don't get air pockets that quickly turn to steam pockets when the engine is running. If your filler is in the radiator, make sure its level is above any water passages in the engine. Jack up the front of the car when filling with water, if needed, to get the filler point higher.

You shouldn't need an oil cooler unless you have gotten carried away with engine compression or you are running longer races than at most hobby tracks.

A shroud helps the fan to pull air through the radiator, but it may have to be fabricated if the engine or radiator are not in stock position. The shroud causes the air pulled by the fan to come through the radiator rather than taking the path of least resistance and going around the radiator. Look at the plastic shroud in the 1979 Mustang II before you fabricate one—it might fit nicely.

New use for old socks. If you run power steering you must slow the pump down or it will blow oil out all day at high RPMs.

To cure his overheating problem, this racer installed a small oil cooler. Note the substantial mounts for both the cooler and the radiator. Note that the oil cooler is mounted in the air path of the radiator. This means warm air exiting the oil cooler enters the radiator—and a loss of cooling efficiency.

A lot of sheetmetal work ducts the air flow everywhere it should be here. No loss in efficiency.

Note the extensive use of tie-wraps to keep everything where it belongs.

This two-piece long hose has a bracket welded in the center. A hose full of water is quite heavy.

Homemade shroud works quite well, as it fits tightly to the radiator. Note upper radiator mount.

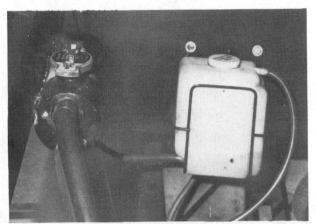
This is a homemade surge tank. It is used because the radiator is lower than the head on the engine.

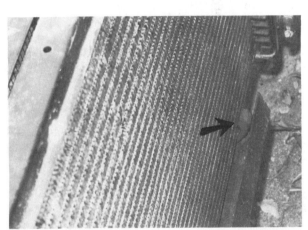
The bottom of the radiator is mounted in rubber to take some of the road shock.

The fan is next to worthless this far away from the radiator with no shroud.

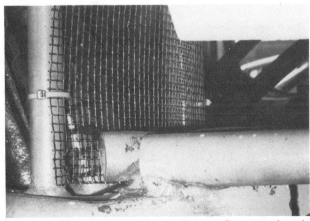
Heavy mesh protects the radiator. Bumper bends before the frame.

I have never had much luck with any of the flexible type of fans—either plastic or steel, but Ford makes seven bladed fans in several diameters and are the same bolt pattern as GM.

Ducting air to the front of the radiator is a big help if you are having overheating problems. This can be done with light gauge sheet metal or aluminum. What you are doing with this is forcing the air to go through the radiator rather than around it. Small holes can be covered using duct tape. Make sure that your design is such that in the event of a light crash, the ducting doesn't wipe out the radiator.

On dirt tracks, you need something to keep the mud from clogging the core. The best item to use is the type of screen seen on many sprinters and modifieds. Make it removable for easy cleaning. If you don't have the facilities to fabricate something like this, then use two or three screens of quarter or half-inch mesh. Don't use fine mesh like window screen. It will clog quickly and even when clean, allows very little air to pass through.

Wiring

Chances are, you've already burned half the wiring while you were gutting the car and putting in the cage. So, rewire the car and use a good grade toggle switch and starter button. Use rubber grommets, or at least tape the wires with electrical tape, where they pass through the firewalls. This prevents wire chafing. Use tie straps to hold wires out of the way so they aren't just hanging. If you have moved the battery to the rear (you should if the rules permit), weld a stud to the frame near the battery and another near the starter. This saves you running a long ground wire as well as a hot lead. If you have an old jumper cable, this can be used for the hot wire by putting the appropriate ends on it. Or, use no. 0 gauge copper arc welding cable for your hot wire. Make sure this cannot chafe anywhere as it passes through both firewalls. Don't run any wires in any area that may be used to jack up the car. This could cause a heck of a trouble shoot if someone breaks a wire with the jack.

Make sure the off-on switch is installed where the driver can easily reach it while belted in the car, and it is clearly marked so that track personnel will know what they are doing in an emergency. Don't use the stock key type of switch unless you plan to spend some time crawling around in the dirt looking for the key some night.

Use at least a 70 amp-hour rated battery, and don't forget about weekly maintenance on it. And be sure the battery is securely tied down within the race car.

Fuel Lines

Many times you can get by with the stock fuel lines. Just make sure they are clipped to the frame so they can't get squeezed shut by someone jacking up the car or by dragging on the track. Don't leave a section of steel line

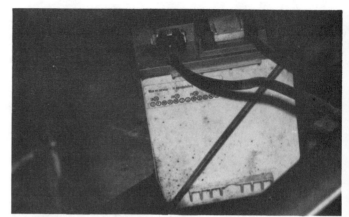
Make sure the battery hold down is secure.

Nobody can drive this one without the key.

Why use a key if you have to wire it fast anyway? Note that when you drill a roll cage tube for a pop rivet like this, you create a weak spot in the tube and it can split or crack the tube eventually.

unsupported as vibration will crack the line. The same goes for brake lines. Those that choose to replace the stock fuel line with neoprene must be careful it is not rubbing against anything or it will wear through. Also be careful of tight radius bends that may kink and shut off the fuel supply. If you choose to run the fuel line through the drivers compartment, it should be run inside a piece of light-wall conduit.

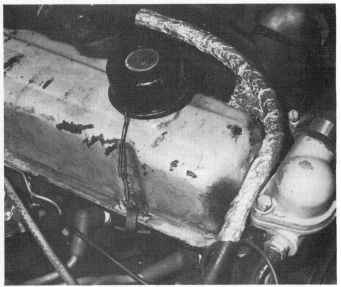

The fuel line has been wrapped to keep the fuel cooler.

A clearly marked fuel shutoff should be within easy reach of the driver and safety crews. Some tracks require that the areas around the fuel shutoff and ignition switch be painted with a Day-Glow type paint, and have "off" and "on" clearly marked. This is not the world's worst idea. Use worm clamps for fuel lines and not the clip type.

Battery has been moved to the rear of the car as it should be and covered to keep acid from leaking in case of a rollover. Fuel line is taped up out of the way.

Use the tin fuel filter and not the plastic kind.

Fuel Tanks and Cells

A fuel cell is a rubber type of bladder filled with a porous foam and enclosed in a tin container. It allows the cell to be dented, bent or twisted without leaking fuel. These come in several sizes with the eight-gallon the most popular for the hobby classes. They are expensive for a hobby car, but if you have ever felt the pain of a burn, the price doesn't seem high at all. Shop around for prices as they vary considerably.

To mount a cell, make a base out of angle iron that runs from chassis rail to chassis rail in the rear, and make a rectangle out of it by welding or bolting in two pieces of angle to just fit the bottom of the cell. The cell can now be tied down using long bolts and angle or strap iron and shorter bolts. Do not use plumber's strap with all the holes in it. Either double-nut the bolts or use self locking nuts so the cell mount can't loosen up from vibration. I have actually seen a fuel cell fall out of a car and drag behind it, so it pays to be careful in mounting this expensive piece of equipment.

If your cell hangs below the frame rails, protect it from a rear end collision with some tubing. Remember that if both you and the person hitting you from behind are on the brakes, the car in the rear will ride under the rear bumper of the front car if they both have stock bumpers.

The track rules may not require a fuel cell. In this case I hope they have good safety inspectors that have guts enough to throw out cars with unsafe tanks or mountings. The mounting of gas tanks is just about the same as described for cells except they should be mounted as far forward as possible in the area behind the rear firewall. The least little nick in a crash usually cracks a seam in a stock type gasoline tank. Try to find a rectangular tank of fairly heavy wall. Many people use what is referred to as "GI Cans" or "Jerry Cans" for fuel containers. These hold five gallons which should be plenty for a hobby feature.

Fuel and ignition switches are clearly marked and mounted within easy reach of the driver.

This is why you never use a stock gas tank in the stock position.

Don't use a tank that is larger than necessary. It will be harder to mount. Don't ever race with a tank in the stock position. This is just asking for a fire.

A stock tank can be moved into the trunk area and strapped to the trunk floor. Use two straps in each direction and cut up an old inner tube to use as cushioning between the tank and the floor so that the vibration will not chafe the tank. Most old tanks are half rusted through to start with, so be careful in your selection.

When modifying a tank, get it steam cleaned inside if it has previously held gasoline. If you don't, it may blow up from any spark. Some people claim that it is enough to fill the tank with water to force out the gas fumes, but any experienced welder can tell you stories about tanks blowing up that were cleaned this way.

Again, we cannot emphasize enough the safety aspect of a fuel cell for your race car. Even if it stretches your budget, it is by far less expensive than the agony—and the cost—or burn treatment.

A nice, clean fuel cell mount using ¼-inch angle iron and J-bolts.

Bottom framework for a fuel cell.

This fuel cell is asking to get hit if someone runs under the rear bumper.

Chapter Six
Transmissions and Rears

Ratios

Whenever one gear is driving another, the amount one turns in relation to the other is called the ratio. If one gear turns three times in order for a second gear to turn twice, their ratio would be 3 to 2. When we discuss ratios in race cars we only use one number as the second number is always one. The ratio in the above example would be 1.5 since it takes one and one-half revolutions of the first gear to turn the second gear one revolution. When someone says they are running a 4.50 rear, it means that the driveshaft must make four and one-half revolutions each time the tires make one revolution.

All of the more expensive race cars use a quick change rear which allows them to run high gear in the transmission. High gear is direct drive, (one turn on the input shaft equals one turn on the output shaft). You will not be able to use this luxury in your car so you must learn to compute both rear end and transmission ratios, which is quite simple.

To find the ratio of any pair of gears, divide the DRIVING gear into the DRIVEN gear. In a rear end, this would be the number of teeth on the pinion gear divided into the number of teeth on the ring gear. Suppose your pinion has 9 teeth and the ring gear has 37 teeth. 37 divided by 9 is 4.11. You have a 4.11 rear end ratio. A transmission is done the same way to find first and second in a three-speed and third in a four-speed.

If you are running high gear in your transmission, your final drive ratio is the same as the rear end ratio. When running other than high, both the transmission ratio and the rear end ratio must be taken into consideration to compute the final drive ratio.

For example, you have a 3.70 rear and are running second gear in a transmission that has a second gear ratio of 1.60. Multiply 3.70 times a 1.60 and you get 5.92. This would be your final gear ratio. The higher the number, the faster the engine will turn to go the same speed. When someone tells you they need more gear, it means a higher number ratio.

Below is an example of the same rear with two different transmissions:

Transmission Gear	Transmission Ratio	Rear End Ratio	Final Ratio
Typical 4-speed			
1st	2.64	3.55	9.37
2nd	1.75	3.55	6.21
3rd	1.33	3.55	4.72
4th	1.00	3.55	3.55
Typical 3-speed			
1st	2.84	3.55	10.08
2nd	1.68	3.55	5.96
3rd	1.00	3.55	3.55

At left, the spider gears in this rear have been welded to lock the rear.

Below, this is a Chevy rear prior to locking. It can't be used for racing in this form.

The only way to know for sure which ratios you have in your rear end and transmission is to count the teeth yourself. People will tell you about tags and stampings and all sorts of ways to find the ratio without getting dirty, but you know for sure if you do it yourself. I've taken too many pieces apart that don't follow the rules.

To check a rear, either pop the cover off or pull the pumpkin. It only takes a few minutes and may save you from over-winding an engine. Mark one tooth on the pinion, rotate it and count the teeth. Do the same with the ring gear. Now you know for sure what you have. If it isn't the one you thought it was, most wrecking yards will let you exchange it (but make sure before you buy).

To check a transmission, put it in the gear you wish to check and pull the side cover. Turn the input shaft to find out which gear is the driving gear and which is the driven gear. Mark a tooth on each as you did with the rear and count teeth.

Remember that if you change tire size, this will effect your gear ratio indirectly. It doesn't actually change the ratio itself, but smaller tires will make the engine turn more RPMs to go the same speed just as large diameter tires will give the effect of less gear.

Locking the Rear

A positraction type of rear may be great for the "streetlight Grand Prix," but they are worthless on the racetrack. You have to lock the two rear wheels so that you can get raction from both. Clean up the rear of grease, crank up the welder so you get plenty of penetration and weld up the spider gears. There is very little chance that you are going to break them loose in a hobby car. If you have a lot of money, you can buy a spool that replaces the spiders, but it is not necessary in your class of racing. Make sure you clean out all the slag so it doesn't float around in the rear when you start running. The gears don't like this at all. If you have a friend with a MIG or TIG welder, they can do this for you without slag. This makes it a lot easier to clean up after welding.

Gearing the Car

The way most beginners figure out which gear to run is by asking around. The problem with this is that many people are very silent about this area of their operation. It may have taken them a lot of time and money to learn what they know about gearing for a particular track. Worse than the person who won't tell you an answer is the one who doesn't tell the truth. Even if you can get the straight scoop from the hot dogs, it may only be of a small amount of help as you probably will not run as fast as they, and, all motors are different. Your engine builder may be of help in selecting a final ratio or at least be able to tell you your RPM range. If you have worked on a car previous to getting your own car, you should have kept your ears open enough that you have a pretty good idea of what is going on in relation to gears.

As a last resort, you will have to figure out a ratio on your own. Following is a method of making a GUESTIMATE. If you have a stop watch or can borrow one, here is how you do it. This is in two steps. First you must calculate the average MPH the cars are running. For this you need to know the lap time of a good car and the size of the track. Lap time is not too hard to measure, but a lot of promoters tend to stretch the truth about the size of their tracks so the MPH figures are higher and thus impress the fans with the speeds. If you are running a quarter-mile track and the cars are turning the track in 15 seconds flat, you plug these numbers into this formula:

$$MPH = \frac{3600 \times \text{length of track}}{\text{Lap time in seconds}}$$

$$MPH = \frac{3600 \times .25}{15}$$

$$MPH = \frac{900}{15}$$

$$MPH = 60$$

The average speed for the track is 60 MPH. You will be going slower than this coming off the corner and going faster as you back off to go into the next corner.

If the horsepower range for your engine is from 4000 to 6000 RPM, then the middle of this would be 5000 RPM. We will use the average RPM to go with average speed in the next formula. (For tire circumference, measure around the outside of the right rear tire at racing air pressure.)

$$\text{Final ratio} = \frac{\text{Avg. RPM} \times \text{Tire circumference}}{\text{Avg. MPH} \times 1056}$$

$$\text{Final Ratio} = \frac{5000 \times 84}{60 \times 1056}$$

$$\text{Final Ratio} = \frac{420,000}{63,360}$$

$$\text{Final Ratio} = 6.63$$

Using the rear and transmission we discussed earlier in this chapter, you can see that your best bet would be the four speed in second gear. This combination gave us a final ratio of 6.21 which is about 40 points less than we would like, but it is better to have too little gear than too much. If you are short of gear, it forces you to work on suspension so the car runs through the corners faster and keeps the RPMs up.

Coolers

You don't need coolers on either the rear end or transmission in a hobby car since the distances you run are short. You should keep the seals in good shape so the oil level doesn't get low. Some racers that have trouble with the seal on the right rear use two seals there instead of one. They cut most of the flange on the second (outside) seal so it can be tapped in right next to the stock seal. If this doesn't stop the leaking, you had better check for a bent axle or housing. If the vent for the rear is on the right housing, run a piece of tubing up in the air about a foot to keep the oil from running out of the vent.

While on the subject of axles, do not switch sides with axles. If you have been using an axle in the left side, do not put it in the right housing. They seem to take a set and snap if switched from side to side. Just mark them on the ends with a little paint if the rear is disassembled.

Types of Rears

Most tracks allow you to run any passenger car rear. Even if this is the case, I would suggest that you use the rear that came with your car until it proves unsuitable. All three major auto manufacturers list quite a few ring and pinion ratios which together with their transmissions will give you a wide range of final ratios from which to chose.

In the GM line there are what is referred to as 10-bolt and 12-bolt rears. This is determined by the number of bolts on the rear cover. The 12-bolt is superior in that it is a heavy duty option and has heavier axles. If you stay with racing, the selection of aftermarket gears for a 12-bolt is better than for a 10-bolt. It may take some searching to find a 12-bolt as they have been snapped up by the drag racers. The 12-bolt Pontiac and Olds rears are different than the Chevy rears. The 1970 and up Camaro and big GM cars have a wider track then the intermediate cars as well as bigger axles.

There is an extensive discussion of the Ford rear in the book, "Race Car Fabrication and Preparation" so I will not repeat that here. Suffice it to say that the information in the chapter on rears is valuable to anyone setting up any rear for the first time.

Check the catalogues at the local speed shop closely and you should be able to find which ratios are available for your rear or what type of rear you must find in the wrecking yard.

Twelve bolt GM rear.

This transmission has been mounted solid with only a thin piece of rubber between the transmission and the crossmember.

Floaters

The only economical way to run a floater rear if your track requires it, is to use a truck rear and wheel adapters that convert it to the modified bolt pattern. If you have the skill and equipment to make your own floater, you can finance your racing by making them for the other racers at your track. The truck rears are massive so you should have no trouble making minimum weight and unless you hit something, nothing should break for years.

In the old days, racers without floaters used to weld tabs to the backing plates that were bent around the flange of the brake drum. In theory, this kept the wheel on when an axle broke. In practice, it wasn't very successful as the axle would break out on the end and the strain would be too great for the tabs and the wheel fell off anyway. If the inside axle breaks, it usually results in the car spinning to the inside of the track, but if the outside axle breaks on the straight while under power it is "Hello walls."

A broken axle without making contact with the fence is not too bad a problem if the axle can be gotten out easily. The backing plate is destroyed and the brake parts are scattered all over the track, but you can go racing that night if you have a spare axle with you. Just block off the brakes to that side, remove the backing plate, put in your spare axle and go racing. Many racers block off the brakes to one side in the rear to get better brake balance anyway.

Chapter Seven
Tires and Wheels

The rules for tires and wheels many times are more liberal at dirt tracks than on blacktop. Many dirt tracks allow the wider tires such as are run on the modifieds. If this is the case at your track, then what you will be using will be used tires and rims that you can purchase as hand-me-downs from the high-buck race teams. If the tire rules are more restrictive, as they should be in a hobby class, then you will be using street-type tires and have a lot less trouble with things breaking (like hubs and axles). In order to be competitive, you have to go as wide as the rules allow with both wheels and tires, which means added expense.

Wheels

There are basically three bolt patterns that are used for hobby cars: 4¾-inch which is used on GM products (except some full sized cars), 4½-inch which is used on Ford, Mopar, and American Motors, and 10¾-inch which is referred to as a Modified pattern. Late Model Sportsman cars have a 5-inch pattern, but most hobby cars don't use this. Don't use four-bolt hubs. They break too easily. If you have studs that are smaller than ½-inch, drill out and replace with this size. If rim width is going to be over 12 inches, you will probably use the modified pattern with an adapter that converts your smaller pattern to the large pattern. These adaptors are available from most places that make wheels. There are also adaptors to convert from six or eight-lug truck rears to the modified pattern.

If your rims are over twelve inches wide, you should consider a truck rear and safety hub on the right front. These wide wheels put a tremendous strain on hubs and bearings. If you paid really good attention in geometry class, you might be able to make your own adaptors, but it certainly isn't a job for an amateur.

Hopefully, your track will have sensible rules on wheels that limit the width. Up to seven or eight inches, you can usually find what you need in a junk yard and even the cost of new wheels is not prohibitive. You should reinforce the centers of stock wheels as they tend to tear out the center secton of the wheel. Check to make sure the center is welded and not just riveted. If this is the case, then weld the center to the outside while you are putting in the gussets

If your rim width will be in the 10 to 12-inch area and you buy wheels, make sure they are for oval track racing. These have double centers and require no extra reinforcing. Do not use the dual bolt pattern wheels that many of the speed shops sell for street use. These seem to crack quite easily. If you are using any type of stock wheel, it is a good idea to give them a visual check each week, or any time you have a slight encounter with anything.

If the bucks are really tight, there are a couple of ways to widen wheels out to as much as 14 inches. The first is to get your local steel fabricator to roll some bands for you in a 13½-inch diameter and the width you want. You then cut the wheel with a torch and weld in the band. It is not as simple as it sounds as you must be constantly checking with a level or you'll be so far out of round that the rim will be worthless. Make sure it is tacked securely in at least four places before you start to finish weld or it will distort from the heat. Weld up the valve stem hole and drill a new one to line up with your tube.

The other method is cheaper, but you pay for it in the long run because it is harder to change tires. In this method, you use two wheels to make one. Simply cut the last inch or two (depending on desired width) off the outside of two rims. Weld them together and cut the center section out of one wheel. These should be definitely reinforced. You'll probably have to put in a new valve stem hole also.

Before you mount tires and tubes on any wheel, go over the inside with a hand file to make sure there are no spots that would chafe a tube. Some people tape over the welded areas on the inside with racers' tape to make sure the rim is smooth.

Since Hobby Stock racing is a contact sport, I suggest running tubes rather than tubeless. This allows you to

These high sidewall tires were mounted on extra wide rims to make them work better.

finish a race even though a rim gets bent. Because of the heat buildup in the tires, passenger tubes don't hold up at some tracks. Ask the guys in your class that are running well what they are using. They may be using racing tubes or bus tubes.

Using Street Tires For Racing

If you must use street tires, there are a couple of tips that might help. The sidewall on a passenger tire is not designed to accept the side loads that you will be giving it. Therefore, it will flex and roll the tire under as you are cornering. At the extreme, this will pull the inner area of the tread off the track and give you less cornering bite. If the rules allow, use a rim that is a couple of inches wider than the width of the tread on the tire. This decreases sidewall flex and allows you to keep more rubber on the track. Air pressure also effects this flex and you will have to run

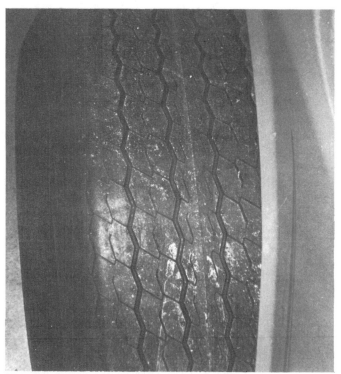

This is a cheater type of D.O.T. tire brand new. Note how little tread depth there is.

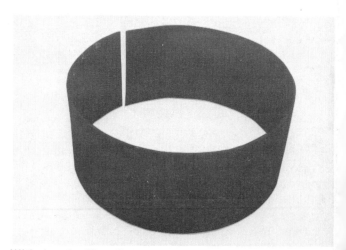

With bands like these, you can make your own wide wheels. You had better be able to read a level though.

See how the excessive camber is causing the tire to wear more on the inside.

This is how you check stagger. Use a thin tape so it will follow the contour of the tire when measuring the tire circumference.

more than you would on the road. The higher the sidewall, the more this flex will be a problem. A crew member closely watching the car in the corner can tell you if the tire is rolling under.

On asphalt, the more tread you wear off a street tire, the better it works. This is because a newer tire with its deep tread will allow the tread rubber itself to flex and does the same thing to each row of rubber that sidewall flex does to the overall tread. The professional classes that use street tires shave the tread down about half way before they use it for racing. Some of the larger tire outlets have tire truing machines that can shave part of the tread off for you. The way most people shave the rubber is to just run the car until the tires start working better.

Many people are using recaps if the track allows it. These are not standard recaps, but use softer rubber than is found in street tires. The softer the rubber, the better a tire will grip the track, but the softer the rubber, the quicker a tire wears out. Soft rubber hides a lot of screwed up suspensions. If you want to pay the bills, you can go a little faster. Since these recaps are racing rubber, the hot dogs that are using them might sell you their used tires. Used racing rubber is still softer than over-the-counter street tires.

Cheater Tires

"Cheater" tires is probably too strong a word. Maybe they should be called rule stretching tires. When tracks in the Midwest came out with rules requiring tires with a

These dual bolt pattern wheels don't seem to hold up very well with wide tires.

Reinforcing rod has been welded around the edge of the wheel. Be sure you don't warp the wheel when welding on the reinforcement.

This is a wide pattern (modified) wheel mounted on a safety hub.

D.O.T. number in an attempt to lower the cost of racing for the late models, it was inevitable that someone would come up with a way to meet the limit of the rules. The Hoosier Street tire was the result. It eliminates the previously mentioned problems with over-the-counter tires because of the design of the sidewall and the small tread depth. The rubber is quite soft as it is in recaps, and, therefore, the tire life is not too long. These and a few other tires such as those designed for street classes at the drag races have filtered down to the Hobby cars in many areas. If these are the tires that are being used at your track, you don't have much of a choice. They are just so much superior to road tires that you can't be competitive without them. But, they cost more money.

I have never seen anyone who was successful with radial tires. They seem to hold well up to a breaking point, but then, the car breaks away without much warning, making it extremely difficult to drive for an expert, let alone a beginner.

Stagger

Stagger is the difference in circumference of the two rear tires. Since you're running a locked rear and always turning left (unless you are a figure eight racer), the right rear tire should be larger in circumference than the left rear and never the reverse. As the track size gets smaller in length, the difference between the two tires gets larger.

You can accomplish this tire stagger in two ways. One way is to simply run less pressure in the left rear than in the right. The other is to use a wider rim on the left than the right. By using a combination of the two, quite a lot of stagger can be obtained with identical tires. Of course you can use tires that are physically different in size.

As tires wear on a race car, the tread will feather to the inside. Never put a feathered tire mounted for the left side on the right side without remounting it for the right. The same is true for right to left. If you do this on asphalt, you'll think the car is on ice. Some people do it on dirt to try to get more side bite, but it doesn't seem to help much unless the track is really sloppy.

Chapter Eight

Suspension

Suspension is a whole world in itself and there are several books on the subject, so here we will try to just give you some basics to go by to get you started.

Most passenger car suspensions are such that with a few modifications, they are quite capable of getting a race car around the track at a pretty good rate of speed. A lot of engineering has gone into them that you can't duplicate without a lot more experience than you have if you're reading this book. The closer you stay to stock, the easier it will be to get parts and fix your driving mistakes. Don't put a left handed framis pin on the thing-a-ma-jig just because someone else has one unless you know what good it is going to do.

As mentioned previously, the stiffer you can make your suspension mounting points, the better off you will be when trying to make the car handle. This allows the suspension to do the work rather than the chassis.

The front suspension will have coil springs or torsion bars. The coils may be mounted on the lower or the upper control arms. It is usually easier to work with the coils on the lower arm, but I have seen a number of cars that worked quite well with coils on the top arm. In all cases, they have gone to some trouble to tie the wedge bolt mount into the front hoop and triangulate the entire assembly. Torsion bars are easy to work with, but they are expensive to experiment with as you would probably need to acquire several rates.

The rear suspension is of four types—leaf springs, or two-, three-, or four point. The two-point uses Chevy pickup truck arms for longitudinal support and a Panhard bar to locate the rear laterally. It has become very popular on asphalt as it transfers a lot of weight dynamically. Since it is not used in any passenger cars, it has to be completely fabricated for a race car and may not be legal under your rules. Make sure that you check before you go to all the trouble.

Four-point is the most popular coil spring suspension type as it is very common on passenger cars such as GM intermediates and some Fords. It consists of two lower trailing arms and two upper trailing arms. The upper arms are mounted at angles so they run from the rear end housing outward toward the chassis. This allows them to control lateral movement as well as longitudinal movement. Some racers add a Panhard bar to this so that the upper arms don't have to work so hard and they can lower the roll center. With a Panhard rod it is possible to disconnect one of the upper arms as both are no longer needed to control lateral movement of the chassis in relation to the rear. This would give you a three-point suspension.

A better way to use three-point suspension is to fabricate a mount for an upper arm so that it can run parallel to the frame rails rather that at an angle like a four-point. The upper arm helps control torque reaction and anti-squat. Many Late Model Sportsman cars use this setup with the upper arm connected to a bracket bolted to the top of the quick change rear. Once again, check the rules to make sure you can make these modifications—or don't ever lift your trunk lid when there is someone around.

Coil springs with two, three, or four-point suspensions can be mounted in front of, on top of, or behind the housing. All of these moutings have been successful, but remember that the further away from the wheels the springs are, the stiffer they have to be because of the leverage factor. So just because some guy is using 300-pound springs in the rear of his car, it doesn't mean that this will be the hot setup for you. If the person selling you your springs doesn't ask you where your springs are mounted in relation to the housing, he may not be able to fully advise you.

Spring and arm mounts for a 2-point suspension with homemade arms.

This bracket is bolted to the housing to convert a 4-point to a 3-point rear suspension.

Above left, the forward mount of a 2-point suspension.

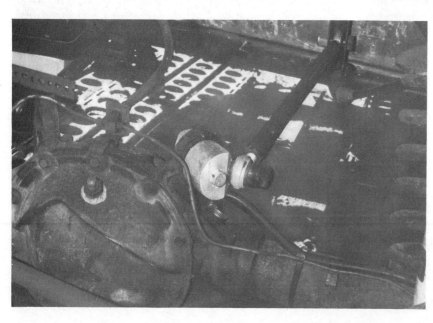
Above left, Chevelle converted to 3-point suspension. Not the typical backyard method, but you may know a machinist.

Coil Springs and Torsion Bars

These are treated together since they are very similar. Assuming a torsion bar of one material, there are only two things which effect its rate: length, and diameter. The longer the bar, the softer it is. The smaller the diameter, the softer it is. The diameter of the bar effects the rate much more than the length. The length of the control arm effects the wheel rate, but in a Hobby Class car, you are tied into the length of arm on the car.

A torsion bar can become overloaded and twist. This can be checked by laying the bar on a flat surface and checking the box at each end before you use the bar. It can then be checked the same way in the event of a bad crash.

Coil spring rate is dependent upon three things: wire diameter, spring diameter and active coils. The larger the wire diameter, the stiffer the spring. The smaller the spring diameter, the stiffer the spring. The less active coils, the stiffer the spring. Active coil means those coils that work as the spring compresses. In a closed wire end of a spring, the first coil does not work. Front springs usually have one closed end and rear springs most of the time have two. The closed end is the flat end of the spring.

Most suspension books give formulas for computing the rate of a coil spring, but it is a lot easier to find a racer or speed shop that has a spring rater. Most speed shops will rate them for you for free if you are, or might become, a regular customer.

If you can get access to a spring rater or make one of your own, you can cut and rate your own springs very cheaply. There is nothing wrong with using used springs

The spring on the left is a variable rate spring. Notice how the coils get closer together toward the bottom. The spring on the right is a linear rate.

Buckets were added to the stock spring mount so the chassis could be lowered without using a very short spring.

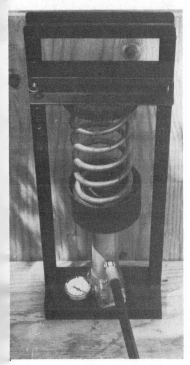
You'll need access to a spring rater if you are going to cut your own springs.

for a race car as long as you don't cut them too short. Leave at least three active coils in the front springs and five in the rear springs.

Stay away from non-linear springs. This means the spring gets progressively stiffer in rate as it is compressed. You can recognize them by the coils being closer together at one end than the other. A linear spring has the same space between the coils for the length of the spring.

Leaf Springs

Leaf springs should be shorter from the centerbolt forward than from the centerbolt to the rear eye. This allows the front of the spring to handle torque reaction and housing location, and the back part to work as the suspension.

The rubber bushings in the front eye should be replaced with something harder. Steel, bronze or aluminum will do.

Solid front mount for a leaf spring.

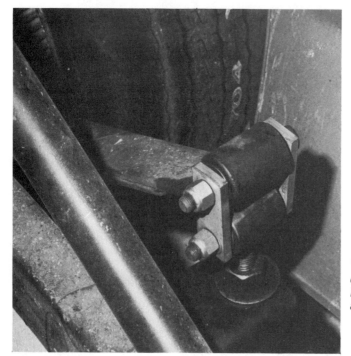

This shackle is going to be working against itself when the car squats coming out of the corner. The spring will lengthen as it straightens out, and you'll find the shackles are too short.

Springs tend to break where they have been heated like this. Don't do it.

Better than using these traction bars is to add half leafs to the front of the spring. This will control spring wrap-up and still allow you to run a soft suspension. Traction bars make the softest springs too stiff because of bind.

This is the right angle for a spring shackle, but be sure it won't bind under normal wheel travel. This one could, because the shackles are so short. Note how wedge can be jacked with this arrangement.

If you use the stock spring mount in the front, when you lower the car to racing ride height, the rear eye will be considerably higher than the front eye. This is good as it gives you more forward bite. You can invert the shackle in the rear so that you don't have to use lowering blocks. Most people stopped using these blocks when they stopped racing flathead engines. Make sure that the shackle pivot on the frame is slightly in front of the pivot point on the spring when the weight of the car is on the spring (see drawing.) Otherwise, the spring will be working against itself as it tries to lengthen under compression.

With the full weight of the car on the springs, check the angle of the U-joint yoke on the pinion. It should be no more than 4 degrees down from vertical. Most racers set it two or three degrees down so that under acceleration it will be straight. You may have to reweld the spring saddles on the rear or shim them to correct this.

When you have the rear end out of the car, you can drill additional centerbolt holes in the saddles to enable you to change the wheelbase of the car if you wish. Shortening the wheelbase this way will also change the characteristics of the springs without getting another set made.

If you are considering using the early Camaro or Nova single leaf spring, be aware that these springs are tapered and much thicker in the center than at the ends. You would be better off to build your own springs than to use these. They give little lateral support to the rear end under cornering.

When making a set of springs, each leaf should have a slightly greater arch than the spring leaf above it. As you pull the centerbolt tight, all of the ends of the springs should be in contact before the centers are touching.

Springs from the front of a Ford Econoline are used often for a starting point. Clean up the rust between leaves and lubricate before assemblying. Clamps are used to keep the leaves aligned and if they are put on too tight, they will make the spring stiffer. You may have to put tape on each side of the clamps to keep them from sliding back and forth.

You can use shackles of differing lengths with several sets of holes if there is no wedge bolt rule at your track. You should not use shackles over 8 inches in length, and they should be made from 3/16-inch thick steel to give adequate lateral support.

Wedge Bolts

Wedge bolts are used to adjust the height of the car as you use different springs. They also allow you to adjust the static weight on each wheel, or what is commonly referred to as "cranking wedge into the car." It's just about the same thing as having a table with one leg longer than the other. If you crank down on the right front wedge bolt, it adds more static weight to the right front and left rear. What the RF gains comes from the LF and what the LR gains comes from the RR. It may not be the best way to adjust a chassis, but Grand National racers use it as a fine tuning tool, and it is the cheapest chassis adjustment you can make.

Remember that wedge is only a fine adjustment, and should never be used as a crutch. You cannot successfully crank wedge into a chassis as a substitute for using the correct spring. If you do, you will have all kinds of handling ills.

Three pieces of tube and a piece of three eighths plate were used to mount this wedge bolt. The tube also gives a place to mount the shocks.

Left, two types of spring retainers. The one on the left uses a cotter key to hold the retainer on the bolt. It has been turned down on the end on a lathe. The retainer on the right is held on by a 3/8-inch bolt. The wedge bolt was drilled and tapped.

Below, when you use a stock frame with a lowered ride height, the rear kick-up area must be notched for increased axle housing travel so the housing doesn't contact the frame. Note how the notch was reinforced.

Since wedge bolts and retainers are cheap, and easy to make, and allow you to experiment with used springs, I don't see why some tracks do not allow them on their Hobby cars. If your track doesn't allow them, then you will have to make several spacers that set up in the spring pockets to get your car adjusted to the height you desire. I have seen some made out of plywood, but they usually don't last long. It would be a cheap way to start until you determine exactly what you need for a more permanent installaton. The finished spacers should be aluminum or steel.

Wedge bolts are made from threaded rod anywhere from 5/8-inch to 1½-inches based upon what you can get free. Most people use one-inch threaded rod. The spring retainer should be ¼-inch thick and have some facility to keep the spring from sliding to the side. 5/16 or ½-inch rod can be bent and welded to the plate to do this. It also helps to get the spring up into the pocket without losing a finger when you change springs.

The retainer has to be fastened to the bolt in some way so that it doesn't fall off everytime you jack up the car. It can be welded to the bolt or if you have access to a lathe, you can turn down the end of the bolt to 5/8 inch and drill a hole for a cotter pin. Another method is to drill and tap a 3/8-inch hole in the end of the wedge bolt for a bolt to hold the retainer on. If you use this method, make sure the bolt bottoms out in the thread so it can't back out. All that remains is to grind the top of the wedge bolt so that it can be turned with a wrench or you can weld a nut of som kind on the top.

Anti-Roll Bars

Most cars come with anti-roll bars of varying sizes. If you nose around the junk yard with a tape measure, you should be able to find several bars of varying diameter with the right length that will fit your car. If the arms are

Put the wedge bolts where you can get at them.

too long, you can always cut them off and weld on new brackets to take the links. The stock rubber mounts on the frame flex too much to let the bar do its job. You can make pillow blocks for mounts, but a cheap, easy way to mount to the frame is by using universal joint U-bolts and pieces of angle iron welded to the frame. Use self locking mounts on the U-bolts and make sure the bar turns freely without being sloppy.

The links from the bar to the A-arm should be made adjustable so that the anti-roll bar can be neutralized each time a chassis adjustment is made. Most of the professional cars use spherical rod bearings for this which makes it quite expensive. You can use most of the hardware (rubber and steel bushings) from the stock links if you replace the long bolt with a threaded rod of the same size and four self locking nuts on each link. A little practice will allow you to squeeze the rubber bushings pretty tight and still have no pre-load in the bar. You'll have to hold the threaded rod with a pair of vise-grips when you tighten the nuts unless you weld a nut to the middle of the rod.

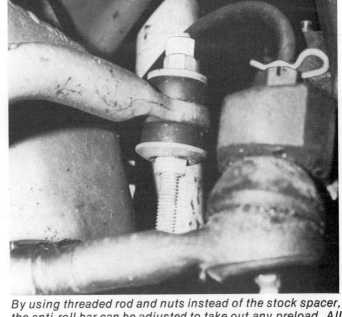
By using threaded rod and nuts instead of the stock spacer, the anti-roll bar can be adjusted to take out any preload. All other parts are stock so it costs next to nothing. Also note that the tie rod has a clip rather than a cotter pin. This makes it quicker to change the tie rod if you are in a hurry.

This lower A-arm has been modified to put a step in it. This allows the whole car to sit lower, but weakens the arm in the event of a crash.

This is a homemade anti-roll bar and pillow block, mounted on the rear suspension.

At right, these are anti-roll bar pillow blocks made from a plastic material similar to teflon called HMW1900. It is available from Cadillac Plastics, and is fairly easy to machine.

Suspension Linkages

Control arms in most passenger cars use rubber bushings so that sound and road shock are not transmitted to he car. This isn't a problem in a race car, and the soft flexible bushings create suspension geometry changes as the car corners. The bushings in the front arms can be replaced with a harder rubber, bronze or steel, but I wouldn't worry about this for a first car. There may be some flex in the rubber, but as a beginning driver, you wouldn't be able to tell the difference anyway.

You can use harder rubber in the rear control arms if you wish, but don't use solid bushings. The rear arms—unlike the front—must travel in two planes and solid bushings will bind the suspension and eventually crack the mounts, If you have a GM or Ford-type three or four-point system, you can find Panhard bars at the local junk yard on several cars or Chevy pickups which can be mounted to your car. Once again, the Panhard bar moves in two planes so you cannot replace its rubber bushings with a solid bushing. If the Panhard bar is too long, you can replace one end with a solid bushing so that it cuts the "slop" in half.

This Panhard bar looks like it was made from spare tubing laying around the shop, but it works.

This Panhard bar mount will probably be pulled out of the chassis because of the distance from the frame, and it is welded to the bottom of the frame, not the side.

This is a homemade panhard bar. Cheap and adjustable. Note the heavy brackets. There is a lot of strain on them.

This is the trailing arm setup for a two-point suspension. Note how the Panhard rod attaches to the arm.

Notice there are several holes in this homemade anti-roll bar arm to change its rate.

Chapter Nine

Driving

Benny Parsons, in his book in driving, said that new drivers are usually of two types. One type is driving over his head from the second hot lap. The second drives within control and tries to gradually go faster. He feels that in the long run, the second type will be more successful. I agree completely. I have never had any use for drivers that tear up equipment driving over their head. Since you are going to have to put your car back together when it gets bent, you probably won't either.

It takes absolutely no skill to push the gas pedal to the floor board and leave it there. It also takes very little skill to drive a race car around a race track fairly fast. It takes a great amount of skill and experience to become a race driver. My own definition of a race driver is: "A person who can consistently run at the limit of the car's capability." Every car, every time it goes on the track has an absolute maximum at which it will lap that track based upon many variables such as track conditions and suspension settings. A race driver, within a few laps, will be running at this speed. The sporty car set calls this "driving at ten tenths." The good driver will also be very consistent in his times from lap to lap. A poorer driver will be very erratic in lap times.

Attitude vs. Budget

If you are racing in a hobby class, chances are that money is a problem most of the time. You can't afford to spend all your funds replacing broken and bent parts on your race car. There is a fine line between being aggressive and being reckless. Only experience can teach you this. If you crash yourself or someone else, it is recklessness. If you move to the front quickly and competively without spinning or crashing, that is aggressiveness.

Usually the track officials will have a discussion with you if you are getting reckless. Some of the other drivers or car owners may discuss it with you from time to time, but as I will explain shortly, their comments may not be too valid. The rule (even though it is unwritten) is that if you get a wheel up along the driver's or passenger's door on a car you are passing, the car that is being passed should give you the lane. For example, on the straightaway, a car moves up so you can see his hood to your left. You should give him room to use the inside groove in the turn. As you gain experience, it will become more of a sound reaction than a sight reaction. Remember, it only takes one ride on the back of a wrecker to have that phase of racing mastered, but it takes a lot of laps on the race track to even learn a little about driving.

Most people that get into oval track racing from drag racing try to treat an oval as two drag races per lap without assigning the correct importance to the cornering

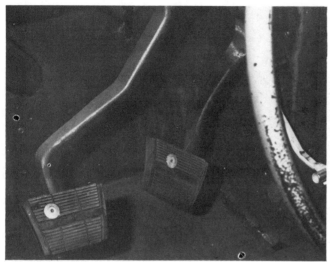
The pedals are bent to the left for comfort, and the rubber is pop-riveted so it can't come off. Good idea.

It will be hard for the driver to monitor the gauges unless he is making right turns.

Here, nothing is marked, and no knob is on the shifter. Needs a little work.

The gauges have been mounted just below the driver's line of sight so he just has to drop his eyes slightly to check oil pressure and water temp. This is good.

of the car. This may be spectacular, but not the fastest way around the track. Ex go-cart drivers and motorcycle riders sometimes have trouble adjusting to the difference in size of their race vehicle. There just is no way you can put a full size car through a four foot space on the race track. Many of these drivers quickly become quite good though, because of their previous experience. If you are not old enough to drive race cars in your state, go-carts, quarter-midgets or motorcycles are a good training ground.

Observation

Use every opportunity to observe other drivers, especially those in the top class at your track. Try to get in places where you can watch the driver's hands. You will find that the better the driver, the less hand movement you will see. The poorer drivers look like they are dealing a hand of five-card stud in each corner. Every time you turn the wheel to correct direction, you are scuffing off speed. *Smoothness* is one of the keys to going fast. Try to observe from different places so that you can find out where the faster cars are backing off and getting back on the power. You won't be able to use the same points with your car, but it will show you what you should be trying to do.

One thing that you will learn by close observation is that the reason some cars seem to have a lot more motor than others is that they are getting on the power earlier and harder than their competition. You can compare this to handicap drag racing where one car is given a slight

At left, a simple way of mounting the window net where it is easy to get to by the driver or the track crew. Below, observation is important in learning the "on-the-track checkers game" of traffic strategy.

head start so that the more powerful car has to really work to catch the car that got the head start.

A saying that I have used for beginning drivers is: "Go in for show, come off for dough." What this means is that it is more important to get back on the gas as soon as possible than to worry about running the car into the corner, deeper than anyone else. Try both methods during warmups while someone is clocking you. The watch tells the truth, as long as someone honest is working it. You will notice that almost all passing on a short track is done because one driver is getting off the corner better than the other. Some top drivers don't even seem to be going fast, because they are so smooth, but are always up front when the race is over.

Dealing With Your Fans and Non Fans

Your pit crew, relatives and friends hopefully will be considered your fans. After you have raced for a while, other people, because of the type of car you drive, its color or any one of a hundred other reasons, will become attracted to you. These people all have one thing in common. They are prejudiced in your favor. In their eyes, you can do no wrong. Keep this in mind when you ask for or are offered an opinion on the night's racing. This is especially important after you have swapped paint with another car during the evening's festivities. Your fans will be telling you it was all the other driver's fault, while his will be telling him it was all your fault. This praise from your fans may be good for your ego, but don't take it very seriously.

A good place for the gauges. The windshield is low enough so the driver can see over it if it gets dirty.

A sturdy driveshaft hoop like this can save the driver a lot of grief if the universal joint breaks.

From the stands or pits, anyone can drive a race car and never make a mistake, but as you found out your first time on the track, this is not true. Don't let your fans or crew get you in a feud with another driver for some minor thing. If you need an objective opinion on something, try to find someone outside this circle of fans.

Summary

The only way to learn how to drive a race car is to drive one. You are going to make mistakes. You are going to bend up equipment. This is all part of the learning process. It is not much different than any other sport. If you want to learn how to play tennis, golf, ping pong or any sport, you must practice. Try, if at all possible, to run more than one track after you get a little experience. If your home track is a quarter-miler, go find a half-mile track that runs your type of car. Don't be afraid to go to another track just because they allow bigger motors. You may be happily surprised.

Wire mesh keeps large objects from coming into the car and breaking your concentration, among other things.

Check the clamp to the right of the rod end. This is to help keep the wheel from being driven back into the driver in the event of a crash.

Know where you are supposed to start "before" you go on the track if you want to get along with the starter.

Chapter Ten
Preparation and Spare Parts

No matter which class you race, you will see certain cars that never seem to break down. This is not an accident. The people who work on these cars "work on these cars." Too many amateur racers are not willing to put in the work on their cars even when they are running once a week. Good preventive maintenance does not cost very much and will save you money in the long run. A race car will not cure itself of its problems by setting on the trailer all week.

Check lists are probably the best way to keep your maintenance procedures up to date. Some racers have a chalkboard on the wall of the garage. On this they list the things that need to be done to the car that week, in addition to the normal things that have to be checked. As these are done, put a check after the item. When they are all completed, then the list can be erased.

If the car has been in a crash, you usually fix that first. If it takes up all of your time, the normal checks will be forgotten for that week. Good crews seem to always have time to fit both into their schedule.

Most people who treat racing seriously, keep a chassis book with the car at all times. As I have said before, it is too easy to forget changes that you make unless you write them down. The following is one type of chassis log. If you have access to a copier, you can make each page like this and it becomes easy to keep track of changes. It is best to set up the car in the garage, because in most pits it is impossible to find a level place to set up the car. If your garage is not much better than these bumpy pits, place marks on the floor so that you can put the wheels in the same place each time.

Date_____ Track_____
Time of day _____ Temperature _____
           Bar Diameter _____
           Arm Length _____

              LF
Type Tire _____
Caster _____
Camber _____
Weight _____
Spring _____
Shock _____
Rim wdth. _____
Circum. _____
Chassis hgt. _____

Toe Out _____

RF
Type Tire _____
Caster _____
Camber _____
Weight _____
Spring _____
Shock _____
Rim wdth. _____
Circum. _____
Chassis hgt. _____

LR
Chassis hgt. _____
Weight _____
Spring _____
Shock _____
Rim _____
Cir. _____

TIRE TEMPERATURES

I	C	O		I	C	O

I	C	O		I	C	O

Ratios
Trans. _____
Rear _____
Final _____

RR
Chassis hgt. _____
Weight _____
Spring _____
Shock _____
Rim _____
Cir. _____

Stagger _____

Times _____
_____
_____
_____
_____

Comments _____
_____
_____
_____

You will have to modify the information depending upon your equipment. For example, if you don't have access to a tire pyrometer, you won't be able to check temperatures of the tires accurately. It's not as good, but feeling the temperature of the tire with the back of your hand will at least allow you to break it down to hot, warm and cold. It is best, however, if you can find access to a pyrometer to accurately check the tire temperatures.

In checking the car between races, some people simply grab a handful of wrenches, (after a few times, you know just which ones), then start at one end of the car and work their way to the other, checking bolts and nuts.

A word about checking bolts: Some people feel that they must make at least a quarter-turn on each item they check or it isn't tight. All this leads to is snapped bolts and stripped nuts. You don't have to turn a bolt to check that it is tight. If you have someone like this on your crew, you had better get him a torque wrench before he breaks everything in sight.

A better way to check over the car is by the function of the components. The following gives a possible check list. You can add or delete for your particular car.

STEERING
Steering wheel nut (Especially if it has a pad on it) _____
Adaptor and bolts _____
Universals _____
Intermediate supports _____
Play in steering box _____
Pitman arm and nut _____
Idler arm _____

FRONT SUSPENSION
Hubs (cracks) _____ (Magnifying glass does wonders)
Bearings _____
Ball joints _____
Inner A-arm mounts _____
Steering arms _____
Anti-roll bar mounts and linkages _____
Shocks _____
Wheels _____
Tires _____
Power steering fluid _____

REAR SUSPENSION
Seals _____
Axles _____
Shocks _____
Panhard bar _____
Trailing arms _____
Shackles _____
Wheels _____
Tires _____
Fluid level _____
Universal joint _____

TRANSMISSION
Linkage _____
Fluid _____
Mount _____
Universal joint _____

BRAKES
Pedal height and softness _____
Fluid _____
Visual for leaks _____

ELECTRICAL
Battery charged _____
Battery hold down _____
Connections _____
Switches _____
Visual for chafed wires _____

COCKPIT
Seat hold downs _____
Seat belt bolts _____
Shoulder harness bolts _____
Crotch strap bolts _____
Head rest _____
Mirror _____
Fire extinguisher hold down _____
Throttle stop _____
Instruments leaking _____
Sheet metal bolts _____

ENGINE
Oil level _____
Oil due for change? _____
Water level _____
Catch can empty _____
Spark plugs _____
Plug wires _____
Coil wires _____
Valves _____
Pulleys _____
Belts _____
Filter tight _____
Hoses _____
Headers _____
Valve covers leaking _____
Carb springs _____
Fuel lines tight _____
Fuel level in cell _____

Run without the dust cover and the grease runs out. Now you get to buy a new bearing.

Wouldn't it be nice to have this in the tow rig? Most guys settle for a "happy box" with all sizes of nuts, bolts and washers altogether.

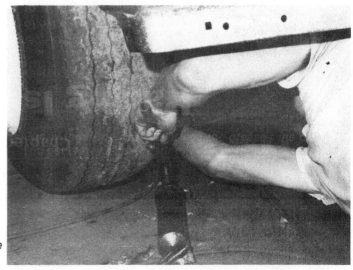
This type of work should be done in the garage, not in the pits.

A box this size should hold most of your spare parts. There are kits on the market to convert these tanks (as seen on the left) to hold compressed air. Get the tank pressure tested first though.

Spare Parts

Most people have heard of "Murphy's Law" (Anything mechanical reserves the right to fail at any time without warning). As regards to spare parts, I have always believed in "O' Leary's Law" which states: "If you have it with you, you won't need it." Since you will probably trailer your car, try to get or build some boxes for the trailer that can be used for spare parts. Even a modest operation should have spare front end parts. If you find a junk car at the local wrecking yard similar in chassis and engine to your race car, you can usually make arrangements to strip it of the things you need. Clean up these parts during your "spare time" and check that ball joints and tie rods are serviceable. Most of the time you won't have available everything you need, but even an old used fuel pump is better than not being able to run because your brand new pump decides it doesn't want to work. Don't carry anything with you that is similar, but does not fit your car. This causes nothing but confusion when you are in a hurry.

The following is a list of spares that a well (super) equipped team will have with them. As I said, you probably will not have all this, but it is nice to dream. I will leave it to you to determine the importance of each item.

ENGINE
Spark plugs
Rotor
Distributor cap
Distributor
Plug wires and wire ends
Coil
Starter with solenoid
Valve cover gaskets
Spare jets
Fan belts
Rocker arm or two
Couple of screw in studs if you use them
Couple of push rods
Engine oil
Couple of wires with alligator clamps in case you have to rig the Off-On switch or starter switch
Head gasket
Intake gasket

COOLING
Stop leak of some type
Spare hoses
Hose clamps
Thermostat(s)
Radiator (real nice item to have)
Radiator cap (or steal off the tow vehicle)

FUEL
Filter
Hose
Pump
Hose clamps

STEERING AND FRONT SUSPENSION
Tie rods (Pre-adjusted saves time)
Steering arms, if separate, and bolts
Idler arm assembly
Intermediate link
Steering box with pittman arm tight
Rag joint if you use it (Not recommended)
Power steering pump and hoses if you use them
Spindles
Hubs
Wheel studs and nuts
Shocks
Upper and lower A-arms (If anything has been changed on the stock arms, do it to the spares also)
Shims or washers
A-arm bolts and nuts
Flex brake lines

TRANSMISSION AND REAR
Transmission and rear oil
Shocks
Seals
Axles
Studs and nuts if different from front
Flex brake line
A spare transmission would put you in the realm of a real pro

GENERAL

Racer's tape
Safety wire
Brake fluid
Bearing grease
Short stub of steel brake lines bent over and soldered (can be used to shut off one brake in an emergency after a crash that wipes out a backing plate)
Permatex
Silicone seal

The next best thing to carrying all this stuff is to be friendly with the guy that does. There is a lot of borrowing that goes on at all levels of racing from the Grand Nationals on down. Just remember to return anything that you borrow, preferably before you leave the track that night if possible. If the word gets around that you are a dead beat, you won't be able to borrow a can of water.

Taken apart and cleaned up, this racer has all his spare front end parts.

Chapter Eleven

Racing Is Fun, But!!!

I have never met anyone with a race car that has ever done it alone. You won't be an exception, so you have to learn how to deal with people that can be a help or hinderance to your racing career. You don't have to be the world's greatest diplomat to survive, but it helps.

Crew

Your crew can consist of your wife or girl friend (husband or boy friend in many cases in recent years), up to a cast of a dozen or more. A crew are those people who actually do necessary work in the racing operation. No matter whether your crew consists of few or many, one person must have the responsibility of the final decisions. Since it's your car, that person will be you. How you make these decisions will many times effect how long your crew members work for you. Since they are not being paid, you cannot treat them as employees. Try to treat your crew as equals in the operation of the car as much as possible. Give praise to your crew when you do well, especially if they are within earshot of your conversations with non-crewmembers. If things go wrong, don't blame them. Since you are in charge, it was your ultimate responsibility even if someone else screwed up.

Try to give each of the people on your crew specific responsibilities based on their abilities. As they gain more experience, their range of responsibilities can be expanded. Some people want to help, but are not too mechanically inclined. You will have to check their work constantly. Sometimes this is best done after everyone has gone home. It saves feelings.

If the budget allows it, you can get T-shirts or jackets for the people that help you. It makes them feel good and helps you find them in the pits easier.

Jackets or Tee-shirts give the crew pride in the car.

Hangers-On

If you run competitively, there will be people who hang around the garage and serve no useful purpose whatsoever. Some of them are so bad that you have to run them out of the place because they not only do nothing, but they get in the way. Others are harmless and can be used to chase parts or coffee. They just want to feel that they are associated with the race car or you as a person. Sometimes you can convert them into useful members of the crew.

Sponsors

If you are lucky enough to find a person or business to help out with the bills, treat them like gold. Make sure his or his company's name is painted on the car in a professional manner so that it shows from the grandstand and in any checkered flag pictures. If the crew gets T-shirts or jackets, so should the sponsor and his kids. Give him an 8 x 10 picture of the car in a frame that he can hang up in a place of his choosing. Get the picture before the car starts to show its battle scars. If you win a race of any kind, get a picture with a checkered flag, and make sure you don't hold the flag over the sponsor's name. Make sure that he gets a copy of this picture also.

Racing is like the Army. You spend a lot of time waiting.

If you can afford it, use a professional for your lettering. If the sponsor's name is neat, he feels better.

You can't pay your crew, but things like this can often be worth more than money.

Officials

The first official that you will come in contact with will probably be the track safety inspector. Some are good and some are incompetent. No matter which type you encounter, they hold in their hands the power to keep your car from going on the race track. Try not to antagonize

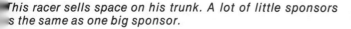
This racer sells space on his trunk. A lot of little sponsors is the same as one big sponsor.

If the official says no, then you don't race.

"Hi there. I'm the track inspector. Mind if I look at your new car?"

them for this reason. I have seen racers give these guys a hard time right from the start and by the time the inspector got through, the racer had two week's worth of work to do to the car before it would pass inspection.

The man with the overall responsibility of running the racing is usually called the pit steward (sometimes the promotor handles this job). If you get caught rough riding or doing stupid things on the track or in the pits, you will get to meet him too. Some tracks require all complaints to be filed through the pit steward. He is also the person who hands out suspensions for whatever track rules spell out this punishment.

You may never meet the starter at your race track, but you had better know how he starts the races. This can be done through observation of what he lets the leaders get away with on starts and restarts. The first time you start on the pole or outside pole, it is not a bad idea to ask the starter what he wants on the start so that you don't get caught napping.

Dealing With The Public

Most tracks open the pit area to the fans when the races are over. This can be a very enjoyable time or pure hell, based upon how you handle it. You should keep in mind that you are an ambassador of racing and that you may be the first race driver that some of the people have ever met. This should keep you from acting like a nut just because you had a bad night. Just think back to the racers you talked to in the pits when you were a fan and how those that you respected treated you. A good guideline is to treat the public the way you would like your wife and kids to be treated under the same circumstances.

The fun part.

Kids can be a lot of fun after the races and all the top drivers treat them like they are very important. This impresses the adults that are around. I have a friend that got a nice sponsorship check because the father was impressed by how he treated his little fan, whether he won or got crashed. Sponsors like racers that get along well with the public.

Summary

Because of the differences in rules for Hobby Class cars from track to track, it has been difficult for me to be as specific in some areas as perhaps you would have liked. There are always people around the track that can help you with these specific problems and are more than glad to do it.

Following is an appendix that contains a vocabulary of terms used in racing, a list of manufacturers of parts for hobby cars, and a list of books that will be of help to you. Books make great presents for your crew members and I'm sure that they will let you read them if you haven't before you give them.

Racers are the greatest people in the world. I hope you are as fortunate as I have been over the last twenty years to meet some of these fine people and to have half the fun with racing that I have had.

APPENDIX A
Common Stock Car Terms

Oversteer — Loose — Assy — Rear end breaks away before front end.

Understeer — Push — Plow — Front end breaks away before rear end.

Stagger — Difference in rear tire circumference. On oval tracks the left rear tire is many times smaller than the right rear.

Wedge — Cross weight — More weight is resting on the LR and RF tires than on the other two. Usually checked by jacking up the car under the center of the rear end and checking how much sooner the right rear tire breaks ground than the left rear.

Caster — Amount top of spindle tilts toward rear or front of car.

Camber — Amount top of tire is tilted in or out.

Toe-out — Front of tires further apart than rear of tires (Toe-in is the opposite).

CG — Center of gravity — Point at which the car would theoretically balance. The geometric center of a car's weight.

Anti-roll bar — Bar with two perpendicular arms connected to frame with arms connected to lower A-arms with adjustable links. Controls body roll.

Sway bar — Panhard bar — Used to control lateral movement of rear end housing.

Trailing arms — Go from chassis to rear housing to control longitudinal movement.

Pyrometer — Used to measure temperature of tire.

Durometer — Used to measure hardness of tire.

Spring rate — Amount of force needed to compress a spring one inch. Measured in pounds per inch.

Wheel rate — Related to spring rate and lever arm of A-arm. Basically, it is the effective rate of a suspension spring at the center of the tire contact patch. Because a spring is placed away from the tire center and is acted upon by a lever arm (A-arm), a leverage factor decreases the spring's effective rate.

Roll center — Point about which the front or rear of the car pivots (rolls).

APPENDIX B
Cage Kit Makers

Beissel Racing Equipment
115 Railroad St.
Concord, MI 49237
(517) 524-8920

Lou Feger's Racing
512 E. Hwy 12
Delano, MN 55328
(800) 328-3618

CSC Racing Products Inc.
163 Bowes Rd.
Box 151
Concord, ON
Canada
(416) 738-2238

Ebeling Engineering
1438 Potrero Ave.
South El Monte, CA 91733
(818) 442-0953

Howe Racing Enterprises
3195 Lyle Rd.
Beaverton, MI 48612

Professional Racer's Emporium (PRE)
1463 E. 223rd St.
Carson, CA 90678
(213) 830-4678

Speedway Engineering
13040 Bradley Ave.
Sylmar, CA 91342
(818) 362-5865

Emanuel Zervakis Ent., Inc.
812 Jefferson Davis Hwy.
Richmond, VA 23324
(804) 232-6729

APPENDIX C
Other Helpful Sources

Racing is not all cutting and welding. A lot of thought goes into building a car. Books sometimes can give you the answer to a problem, but even if they don't, they get you thinking. The initial cost seems high (what doesn't), but if you can find one idea from a book that saves you some money or time, it has paid for itself. Drop hints in the right places and you can get the right books for birthday or Christmas presents.

The following is a short book review of some publications that will be of help in becoming a serious racer.

RACE CAR FABRICATION AND PREPARATION— Steve Smith Autosports. The best eight bucks a beginner can spend. 167 pages of circle track information for anyone running anything from a Street Stock to a Grand National. Full of pictures and tables. See ordering information at rear of this book. Full of pictures and tables.

RACING ENGINE PREPARATION— Steve Smith Autosports. Written by Waddell Wilson, a top engine builder on the Grand National circuit. You don't have to get this carried away for a Hobby Stock motor, but lots of helpful hints and good theory.

STOCK CAR DRIVING TECHNIQUES— Steve Smith Autosports. Written by GN driver Benny Parsons. If you don't know who he is, then you are not too interested i racing. Will not do you much good until you have driven race car a few times, but then it will really start to mak sense.

CHILTON'S AUTO REPAIR MANUAL and **MOTOR' AUTO REPAIR MANUAL**— Basic information on repair o all types of cars. The current edition covers the last s years, so check that the edition you get covers the car yo have. Lots of good general information on brake transmissions, rears, etc. Almost all libraries have copies they haven't been ripped off by the do-it-yourselfers. Cu rent editions are carried in most book stores and Sear Older editions can be found in used book stores.

ADVANCED RACE CAR SUSPENSION DEVELOPMEN — Steve Smith Autosports. Don't be afraid of the title. Th book starts with the basics and is only as complicated a you want to make it. Good explanation of chass dynamics.

Check out the book sections in your local speed sho book stores and public library for books on specif engines or carburetors you plan to use.